AERIAL INTELLIGENCE

Exploring UAV Applications Through Data Acquisition and Analysis

GARVIT PANDYA

notionpress.com

INDIA • SINGAPORE • MALAYSIA

ISBN

Hardcase 979-8-90451-323-8
Paperback 979-8-90392-742-5

Contents

PART II: Drone Data and Mission Planning

PART III: Industrial and Commercial Applications of UAV Technology

PART IV: Future Technologies and Emerging Applications

Acknowledgement

I extend my sincere gratitude to the readers of my first book, *Basics of Unmanned Aerial Vehicles*, for their valuable feedback and reviews. Special thanks to universities, colleges, RPTOs, training organizations, and state/central government institutions in India for adopting it as a textbook and reference. This widespread acceptance inspired this second volume.

My deepest appreciation goes to my family: my father, who meticulously reviewed every line to ensure simplicity and clarity for students; my mother, for her unwavering care and understanding; and my wife, for creating a distraction-free environment during long writing hours and for her invaluable support in graphics.

I am grateful to the late Mr. Dilip Mehta, Sir, for the knowledge and skills that shaped my writing journey, and to the late Mr. Yogendra Jahagirdar, Sir, for guiding my path in the UAV industry.

Heartfelt thanks to my well-wishers for their silent support, and to my employees, interns, customers, and personal/professional connections who enriched my life during this writing journey.

With gratitude and best wishes for your continued success.
Garvit Pandya

Preface

The 21st century has sparked a "UAV Revolution," propelled by advances in communication, the internet, GPS, electronic chips, and more. UAVs, which date back to the early 1900s, are evolving rapidly to transform how we observe, analyse, and interact with the world.

This book delivers a structured, application-oriented guide to UAV technology, tailored to missions from commercial to defense needs. Organized into four parts, it progresses from fundamentals to advanced uses:

- Part I: UAV basics, types, components, and payload systems.
- Part II: Mission planning, data acquisition, processing, and software tools.
- Part III: Commercial and industrial applications, covering traditional challenges, UAV solutions, data capture, and key deliverables.
- Part IV: Emerging uses in defense, AI integration, and future industry scopes.

Intended for students and aspiring professionals, this book equips you to apply UAVs effectively in real-world scenarios.

"Drones: Where curiosity takes flight and careers soar."

Abbreviations

AF – Auto Focus
AGM – Air to Ground Missile
AGL – Above Ground Level
AI – Artificial Intelligence
ASM – Air to Surface Missile
AUW – All-up weight
BLDC – Brushless DC Motor
BVLOS – Beyond Visual Line of Sight
CCTV – Closed Circuit Television
CFRP – Carbon Fiber Reinforced Plastics
CMOS – Complementary Metal-Oxide-Semiconductor
COTS – Commercial off-the-shelf
DARPA – Defense Advanced Research Projects Agency
DEM – Digital Elevation Model
DGCA – Directorate General of Civil Aviation
DHM – Digital Height Model
DSLR – Digital Single Lens Reflex
DSM – Digital Surface Model
DTM – Digital Terrain Model
EO/IR – Electro-Optical Infra-Red
ESC – Electronic Speed Controller
EFP – Explosively Formed Penetrator
FCS – Flight Control System

FIDA – Federation of International Drone Soccer Association

FOV – Field of View

FPV – First Person View

GCP – Ground Control Points

GCS – Ground Control Station/System

GIS – Geographic Information System

GCU – Gimbal Control Unit

GNDVI – Green Normalized Difference Vegetation Index

GNSS – Global Navigation Satellite System

GPS – Ground Positioning System

GSD – Ground Sampling Distance

GW – Giga Watts

HALE – High Altitude Long Endurance

HD – High Definition

HDMI – High – Definition Multimedia Interface

HFOV – Horizontal Field of View

HUD – Head-up Display

IC – Internal Combustion

IMU – Inertial Measurement Unit

INS – Inertial Navigation System

ISR – Intelligence, Surveillance, and Reconnaissance

KML – Keyhole Markup Language

LCD – Liquid Crystal Display

LED – Light Emitting Diode

LIDAR – Light Detection and Ranging

LiPo – Lithium Polymer

Li-Ion – Lithium Ion

LOS – Line of Sight

MALE – Medium Altitude Long Endurance
MAV – Micro Aerial Vehicle
MP – Mega Pixels or Million Pixels
MW – Mega Watts
MTOW – Maximum Takeoff Weight
NAV – Nano Aerial Vehicle
NDVI – Normalized Difference Vegetation Index
NIR – Near Infrared
OLED – Organic Light Emitting Diode
PPK – Post-Processed Kinematic
PTZ – Pan, Tilt, and Zoom
PU – Polyurethane
QGC(S) – Q Ground Control (Station)
RC – Remote Controller/Remotely Controlled
RF – Radio Frequency
RGB – Red, Green, and Blue (Visible Lights)
RPAS – Remotely Piloted Aircraft System
RTF – Ready to Fly
RTH – Return to Home
RTK – Real Time Kinematics
RTL – Return to Launch/Land
SAAS – Software as a Service
SAVI – Soil-Adjusted Vegetation Index
SLAM – Simultaneous Localization and Mapping
SOP – Standard Operating Procedure
UAS – Unmanned Aircraft System
UAV – Unmanned Aerial Vehicles
UCAV – Unmanned Combat Aerial Vehicles
USAF – United States Air Force

VFOV – Vertical Field of View

VTOL – Vertical Takeoff and Landing

VTX – Video Transmitter

VRX – Video Receiver

WW1 – World War 1

WW2 – World War 2

PART I

FOUNDATIONS OF UAV TECHNOLOGY

CHAPTER 01

Fundamentals of Unmanned Aerial Vehicles

Unmanned Aerial Vehicles, or drones, are machines designed to fly without any humans on board. During the initial period of controlled drone flights, pilots used a remote controller to operate the drones from the ground. These initial vehicles were fixed-wing UAVs, often considered RC (Remote-Controlled) models, primarily due to their lack of self-stabilizing and autonomous flight capabilities. Later on, UAVs evolved into many different types and were designed over many years for a variety of applications, spanning the commercial and defense sectors.

These vehicles were built with electronics that allow them to fly with a pre-programmed mission and pre-installed control inputs along the mission, which is monitored from a ground station to ensure their successful operation. The long-range UAVs have tremendous capabilities in terms of range and endurance. These UAVs operate via satellite communication systems to monitor operations and control the UAV as and when required.

The latest edition of such drones may operate over a 4G, 5G, or satellite communication network. These drones have limited range and endurance. With 5G communications, the range of drone operations has been extended; however, due to reduced endurance, the benefit of 5G communication remains limited to short range.

1.1 Types of UAVs

The types of UAVs are defined and decided based on their structural appearances. There are five types of UAVs, of which four are commonly used in professional, commercial, and defense applications.

a. Fixed-wing UAV
b. Multirotor UAV
c. VTOL UAV
d. Helicopter UAV
e. Ornithopter UAV

A. Fixed-wing UAV

Fixed-wing drones are aircraft-shaped or scaled models of conventional aircraft, with the wings fixed perpendicular to the central body to generate lift. Fixed-wing UAVs can be operated without a human pilot on board, using a remote control and GCS (Ground Control Station), and can also operate autonomously. Fixed-wing UAVs are also named as fixed-wing UAS (Unmanned Aerial System) and fixed-wing RPAS (Remotely Piloted Aerial System).

Image 1.1 – Fixed-Wing UAV Model[1]

B. Multirotor UAV Systems

A multirotor UAV is a vehicle designed with multiple motors or rotors fixed on the arms of the vehicle in horizontal positions to generate lift and thrust and to control its operation and motions. The number of motors ranges from 2 to 8, or even more, depending on requirements and the set configuration. So, as the number of motors changes, the drone's name also changes. The one with four motor configurations is known as a quadcopter or quadrotor. The six-motor configuration is known as a hexacopter. The reason behind the word "Multirotor" is to express and address all the multiple rotor-mounted platforms and vehicles as the same physics of flying and operation applies to each one of them.

Image 1.2 – A Multirotor System[2]

C. Hybrid VTOL UAV Systems

VTOL stands for Vertical Take-Off and Landing. As the name suggests, it is a type of drone that takes off vertically and lands vertically. It is necessary not to confuse it with multirotor systems. It is quite correct that a multirotor system also takes off and lands vertically; however, VTOL drones combine features of fixed-wing drones and multirotor systems to include advantages of the both. So, a VTOL drone will take off vertically like a multirotor and transit forward like a fixed-wing UAV, removing the requirement of runway and increase the endurance.

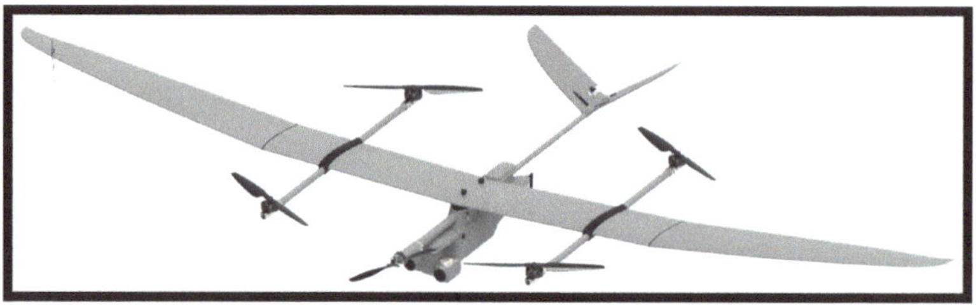

Image 1.3 – A VTOL UAV[3]

D. Helicopter UAV Systems

A helicopter UAV system operates like a traditional helicopter; however, it differs from fixed-wing drones as it can hover. The technical parameters, aerodynamics, and flight operations are entirely different from those of an RC aircraft. A helicopter drone is a type of drone whose operation falls between the multirotor system and fixed-wing systems, and it is more typical than either. Apart from the hovering advantage, helicopters can take off and land vertically.

Image 1.4 – A Cam copter Helicopter Drone[4]

E. Ornithopter UAV

Ornithopter UAVs are designed with a shape similar to that of birds; during operations, their parts, or wings, flap like a real bird, mimicking the natural process of flight, often called a flapping-wing vehicle. This bionic bird-type UAV is best suited for surveillance, mainly because of its structural appearance and the fact that features such as a camera are integrated into the eye. These UAVs are designed using operational references from many real birds' structures and operations. The major benefit and threat are that it can be camouflaged as a real bird and cannot be identified as a drone by the human eye or radar.

Image 1.5 – Ornithopter UAV[5]

1.2 UAV Classification (International Standard)

UAVs are flying objects, and the sky is the limit for UAV operations. To address safety, security, and quality concerns, governments in different countries took the necessary steps to regularize drone operations. The regulatory standards defined by these countries differ from one another. However, a certain level of common standards is present, and as a result, the classification explained below is in accordance with the international standard.[6]

1. Small UAVs

a. Nano and Micro UAVs

As the name suggests, nano drones are the size of a mosquito or a bee. To design such a drone, it is necessary that the sub-parts, viz., camera, propulsion system, and flight controllers, are small enough to be considered collectively in the nano category when fully developed. This drone category is small but effective for short-range surveillance.

b. Micro UAVs – Micro Aerial Vehicles

MAVs are considered to be not more than 150mm in size, maximum, and capable of flying only inside any enclosed premises.

Note – An international MAV category should not be confused with the micro-drone category as defined by the DGCA in India.

c. Mini UAVs

The mini drone is categorized under the MTOW condition of 20 Kg. maximum. This type includes fixed-wing, hexacopter, octacopter, and helicopter UAVs.

2. Medium Range UAV

Medium range UAV - Medium range or tactical UAV with a range of 100 to 300 km. These air vehicles are smaller and operated within simpler systems.

3. MALE UAV

MALE (Medium Altitude Long Endurance) UAVs fly for around 24 hours at an altitude up to 15000 meters. The range for a MALE drone may exceed 500 km.

4. HALE UAVs

HALE (High Altitude Long Endurance) UAVs fly above 15000 meters of altitude for more than 24 hours. These drones are used for extremely long-range surveillance purposes. This also includes drone operations across the border and cross-border issues.

1.3 UAV Classification as Per Indian Standard

Drones, or UAVs, are designed according to application requirements. In other words, different use cases require different types of drones, and not all use cases can be covered with a single micro drone or a small drone. This makes each drone different from the other. Brief information on these drones, by weight category as per Indian drone regulations, is given below for better understanding.

For Indian Standards, the following categories have been decided by DGCA (Directorate General of Civil Aviation).[7]

A. **Nano Drones** – Nano drones are categorized as a drone with MTOW less than or equal to 250 grams. These are advanced systems that take drone applications to a new level, primarily for defense purposes. These applications mainly include surveillance in secret missions inside the forest and border security. These drones are quite small and fast, making them difficult to identify visually and even by radar.

Image 1.6 – A Black Hornet Nano UAV[8]

B. **Micro Drones –** Micro drones should weigh between 250 grams and 2 Kg as MTOW. MAVs can be a tri-copter, quadcopter, ornithopter, or fixed-wing. This type of miniaturized drone is mostly used for consumer applications such as land mapping, various surveys, and inspections, carried out with the help of various available sensors.

Image 1.7 – Parrot Anafi UKR Drone [9]

C. **Small Drones –** Drones weighing between 2 kg and 25 Kg of MTOW are considered small drones. Drones in the small category include quadcopters, hexacopters, octacopters, helicopters, and fixed-wing aircraft. This category includes drones used for applications such as surveillance, film photography, pesticide spraying in agriculture, and delivery.

Image 1.8 – ALTI UAS Transition[10]
(Credit – ALTI UAS)

D. **Medium Drones –** Medium drones weigh between 25 kg and 150 Kg of MTOW. These drones can be octocopters, helicopters, or fixed-wing, in most cases. With a capacity of more than 25 kg, these category drones are quite large and can be used for anything from cargo delivery to firefighting. These drones are heavy, of course, because of the payload. However, as the payload size increases, the vehicle's frame size and weight also increase, along with the high-end, heavy propulsion system and power source.

Image 1.9 – A Velos V3 Helicopter UAV[11]

E. **Large Drones –** Large drones have an MTOW above 150 kg. Octocopters, helicopters, and fixed-wing aircraft can be developed with MTOW above 150kg, depending on the application. Such UAVs are large, sometimes even the size of a fighter aircraft. Military drones will include some as surveillance drones and some as unmanned combat/fighter aircraft that carry guided missiles/bombs and other warfare payloads.

Image 1.10 – RQ70 Stealth UAV[12]

1.4 UAV Sub-Systems and Components

Drones are made up of a collection of parts. These parts are divided into two segments. One includes a complete structure of a UAV, often called the frame. These body parts are made of composites, aluminum, and other materials, as per the drone's design and requirements. These parts comprise frames, wings, arms, landing gears, etc. The second one includes all the electronic, electrical, and mechanical components. These components are what let a UAV fly. The brief explanation of all the key components is provided below.

1. Frame or Body

A frame or body is the main physical structure of the drone, with enough strength and capability to support all the electronics and other components. These frames are designed per the requirements and then developed using a single or multiple materials, as required by the strength requirements. These frames consist of a central body, arms, and sometimes, landing gears.

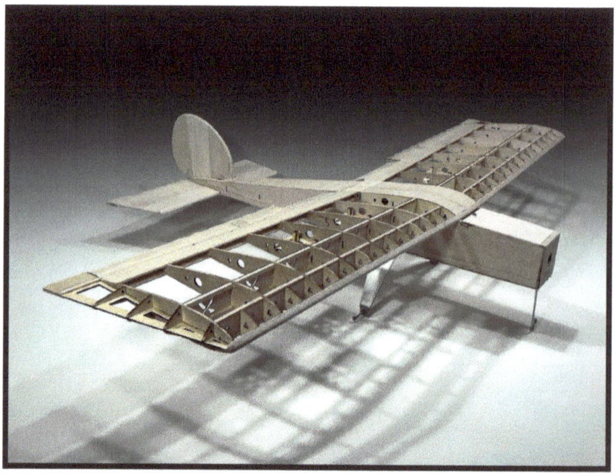

Image 1.11 – Frame of a Fixed-Wing UAV[13]

2. Propulsion System

The propulsion system of a drone consists of a mechanism that generates thrust to move the object in a specific direction. This propulsion system can be electric or engine-based. Brushless DC motors are mostly considered an electric-based propulsion system. Whereas, engine-based propulsion systems include nitro engines, gasoline engines, jet engines, etc. In electric propulsion systems, BLDC motors are always paired with an ESC (Electronic Speed Controller).

Image 1.12 – A BLDC Motor[14] and 2 – Stroke Engine[15]

3. Battery

On any UAV, a battery is a critical component that powers the entire system. A battery in any drone can power the following: BLDC motor operation, control surface movement, engine ignition, camera triggering and gimbal operation, flight controller operation, and other electronics. Normally, drones are powered using LiPo (Lithium Polymer) or Li-Ion (Lithium Ion) batteries.

4. Main Brain Unit/Flight Controller Board

The flight controller system is considered the brain of any UAV because it automatically controls the drone's operations and maneuvers based on the pre-planned mission entered into the system. Flight controllers are connected to all the drone's avionics, allowing them to take command of the entire operation.

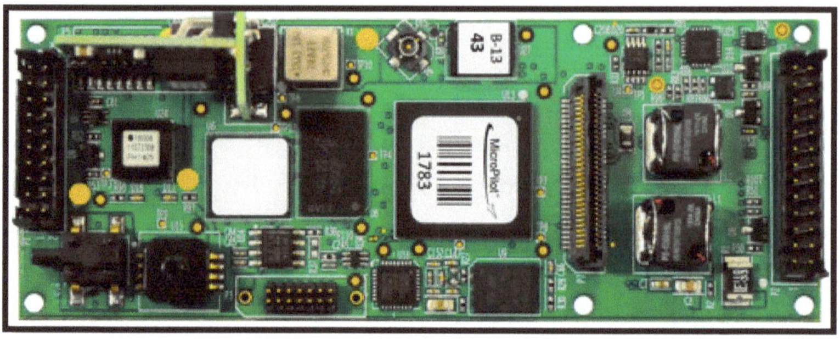

Image 1.13 – Micropilot Flight Controller Board[16]

Along with the flight controller systems, there is a GPS connected to it. This is useful for accurate drone positioning, which can be used for hovering a multirotor system, following the path provided in a pre-planned mission, autonomous operations, return-to-home, etc.

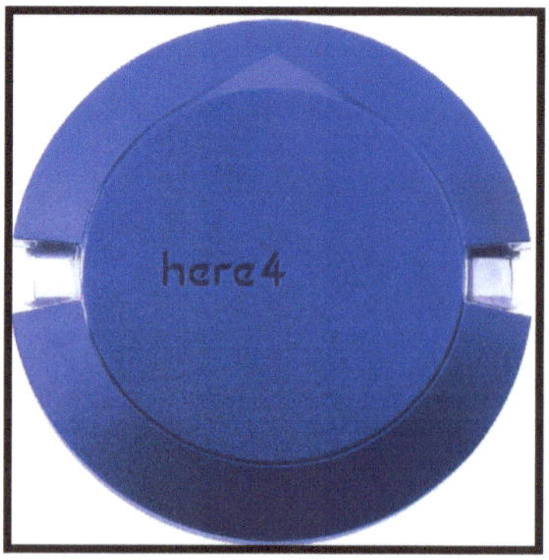

Image 1.14 – A GPS Used on a Drone[17]

5. Propellers

Propellers are the blades that are attached to the engine or motor and generate lift/thrust. This generation of lift/thrust depends on the motors' positions. The propellers are identical to the wings of a fixed-wing UAV, both fulfilling the main purpose of generating lift.

6. Communication systems

UAVs are operated manually and autonomously, and in both cases, communication systems are essential for either controlling manual flight or receiving data from the drone on its position and other important parameters in autonomous flights. Manual flights are conducted using a remote controller. Whereas long-range telemetry systems, 4G, 5G, etc., are used for the autonomous flight operation of a UAV.

Image 1.15 – A Handheld GCS Useful for the Military[18]

Apart from the above systems and sub-systems, there are other parts of UAVs, such as the power distribution board or power management unit, battery chargers, servo motors, payloads, etc. The payload is covered in this book as a separate chapter.

Overall, UAVs are designed and developed to cover almost all the applications that are difficult, risky, and time-consuming when performed manually.

(For the detailed explanation about UAVs, it is recommended to refer the book – Basics of Unmanned Aerial Vehicles[19])

Points to Remember:-

- There is a variety of UAVs depending on their usage
- UAVs can be of different types and different sizes
- Calculations during the design and development of a UAV play an important role in efficiently achieving the desired application.

CHAPTER

02

Application Based UAV Payloads and Sensor Systems

Payloads are the load that a drone can carry to execute certain missions and applications by operating it during drone flight – manually or autonomously. UAVs are composed of a variety of parts, viz., frame, canopy, landing gear, propulsion system, communication systems, control board electronics, etc., which are essential for a drone to fly autonomously or manually. When a UAV is developed, these parts are selected and are permanent on the drone; they may need to be replaced, but will never be removed if one wants to fly the drone. Payloads, on the other hand, are external, separate components of a UAV. A UAV will not stop flying if the payload is removed. On the contrary, it will fly longer because adding any payload increases the UAV system's weight. These payloads are typically in the form of a camera, an explosive warhead, a logistics box, sensors such as LIDAR, a robotic arm, a water tank, medical equipment, a food box, etc. Whether to choose a pizza box, a night-vision camera, or a missile depends entirely on the application for which it is intended, in accordance with the UAV's capabilities and the purpose for which it is designed and developed.

There are applications where the camera or sensor payload may remain the same. For example, in highway surveys, a mapping camera is used. The same camera is useful during a land survey. Both applications are different. The desired results are separate, so the data processing will also vary. This results in completely different reports because the requirements for the two use cases are not similar.

These cameras or sensors, when attached directly to a drone, are always affected by vibrations, resulting in distorted or blurred images. These types of blurred photographs are of no use. To compensate for vibrations, 2-axis or 3-axis gimbals are used. Sometimes, when gimbals are not necessary, dampers can be used between the drone and the payload.

Some essential payloads for current UAV technology are listed here:-

a. Gimbal stabilization system
b. RGB camera system
c. Zoom camera system
d. LIDAR sensor
e. Thermal imaging camera
f. Multispectral imaging sensor
g. Hyperspectral imaging sensor
h. Agricultural spraying system
i. Aerial Delivery system
j. Defense payload system

2.1 Gimbal Stabilization System

A gimbal is a support and camera-mounting platform that allows an object to rotate about 2 or 3 axes. Gimbals are powered by 2 or 3 brushless motors for 2 or 3 axes of rotation, respectively. These 3 axes are pan (yaw), roll, and tilt (pitch). The rotation operation can be performed using a remote control or a ground control station. It consists of a GCU (Gimbal Control Unit), which controls motor operation based on input for angular change. IMU sensors are also part of its control system, which allows the GCU to control the motors and stabilize the system at the desired position.

For drone-based photography, it is recommended to mount the camera on a gimbal, as it provides stability in the air, dampens vibration, and keeps the camera at a fixed position. This results in better image quality and smooth videography. The weight of a gimbal ranges from a few grams to a few kilograms, depending on the camera to be mounted. Not all gimbals support all types of cameras; they support only the specific cameras for which they are designed.

Image 2.1 – Gremsy Gimbal for Sony Alpha 6000 Camera[20]

Gimbals are designed and developed with and without a camera, and an integrated camera system. When a gimbal is built without a camera, it only supports a few cameras it is designed for. Also, generic gimbals are developed to allow any camera to be mounted. However, this comes with certain limitations, such as the absence of an image capture trigger, and it can support cameras with certain dimensions and weight.

2.2 RGB Camera System

RGB cameras are regular cameras with red, green, and blue color models. The simplest example is a smartphone that captures an RGB image. Applications like mapping and surveying do require a crystal-clear photograph from a visible-light camera (RGB Camera). However, the quality of these cameras is equivalent to or better than DSLR-type cameras. Some commercial off-the-shelf cameras are compatible with mapping and surveying. Additionally, there are camera companies that design and develop cameras for various purposes, but meant to be mounted on a drone. They aim to keep the camera's quality comparable to DSLRs or better while reducing weight as much as possible to keep it light. This can be done by removing all the unnecessary parts of a DSLR camera. Because a mapping camera for a drone does not need the same facilities as an actual DSLR, like flash, display, and certain other buttons are not required. This way, a dedicatedly manufactured mapping camera is lighter than a COTS DSLR camera.

a. Sony α

Sony α is Sony's new mirrorless camera series, which includes additional features alongside the common features of the Nex series cameras. This RGB camera is quite preferred for a drone-based survey when it comes to a self-developed or customized drone, as it has the fastest AF, which captures sharp images and is capable of capturing images remotely, which is an important requirement for a drone user.

Key Features:-

- 24.3-megapixel APS-C CMOS sensor
- Bionz X image processor
- Hybrid AF system with 25 contrast-detect and 179 phase-detect points
- Built-in flash + multi-Interface Shoe
- 11 fps continuous shooting with subject-tracking
- 3-inch tilting LCD with 921,600 dots

- OLED electronic viewfinder with 1.44M dots
- Diffraction correction, area-specific noise reduction, and detail reproduction technology
- Full HD video recording at 1080/60p and 24p; clean HDMI output
- Wi-Fi with NFC capability and downloadable apps
- Image size – large – 6000x3376
- Weight – 350gms
- Dimensions – 127 x 67 x 45mm
- This camera is not waterproof

Image 2.2 – Sony Alpha 6000[21]

Sony a6000 can be controlled remotely from the ground control station for camera triggering, which is done through the autopilot system.

b. ADTI Cameras

ADTI is a technology company engaged in UAS payload systems design and development. ADTI develops a camera dedicated to the UAV industry. There are various survey-grade cameras ranging from 20MP to 61MP, and some gimballed cameras as well.

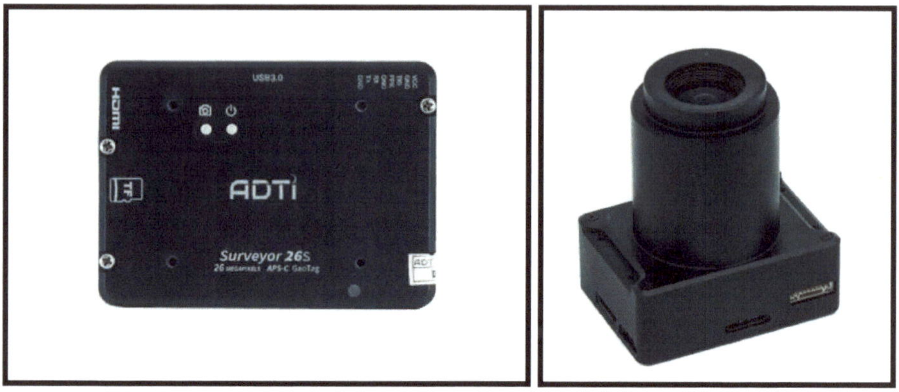

Image 2.3 – ADTI 26S Survey Camera[22]

Key Features include:-

- 26MP APS-C CMOS Sensor
- Mechanical shutter with fixed focus lens
- Built-in multi-interface hot shoe output for PPK
- Image size – 6192x4128
- Weight – 120g with lens (Aluminium case)
- Dimension – 58 x 58 x 23 mm

Drone-Based Applications:-

- Land mapping
- Forest area survey
- Highway surveys
- Any large area mapping

2.3 Zoom Camera System

The above-mentioned RGB cameras for mapping applications do not have a zoom feature. Hence, it becomes difficult to capture images and video of certain critical areas. For example, a UAV is flying over a national highway at an altitude of 100 meters when an accident occurs. To cover the same, the drone needs a camera that can accurately capture images or video. To do this, a zoom camera is effective, as only the camera needs to be controlled, whereas with a mapping RGB camera, the operator will need to take control of the drone to reduce altitude to capture images in an accident zone.

Zoom cameras are designed to capture videos at different zoom levels and orientations. Because of this, the Zoom camera normally comes with a gimbal (mostly 3-axis).

Image 2.4 – 10X Zoom Camera[23]

The image above shows a 3-axis, 10x optical zoom camera system. It has a 1/3 CMOS sensor and 4 MP resolution. In addition to optical zoom, it supports 8x digital zoom. It has an in-built gimbal system, so there is no requirement for an external gimbal. All three axes are controlled with three BLDC motors. The weight of such a gimbal ranges from 300 grams to 600grams.

Drone-Based Applications:-

- Surveillance
- Utility inspection
- Security
- Crowd monitoring

2.4 LIDAR Sensor

LIDAR is a "Light Detection and Ranging" Sensor that creates a 3D model of the surveyed area. The sensor's operation is based on the emission of laser light and the pulses that return to it. The time taken by the pulses to return is calculated, and the variation in the distance is measured. LIDARs for small commercial UAVs weigh somewhere between 500 grams and 2 kg, depending on the specifications. Most of the LIDARs have a horizontal FOV of 360°.

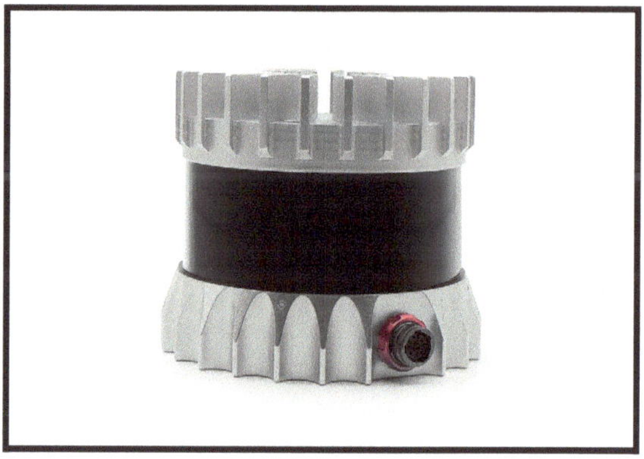

Image 2.5 – Ouster LIDAR[24]

LIDAR is an advanced sensor technology that provides more accurate results than RGB cameras. Because LIDAR sensors can send a pulse even through small gaps, for example, between tree leaves, and collect ground points. This helps generate 3D point clouds with more points, resulting in higher accuracy.

Drone-Based Applications:-

- 3D point cloud generation of an area
- Forest mapping
- Terrain mapping
- SLAM for mining

2.5 Thermal Imaging Camera

Thermal cameras are infrared cameras that detect temperature and allow us to identify the temperature of multiple objects in a single picture, as all objects emit infrared energy according to their thermal properties. With temperature variations, it becomes easier to identify objects, their characteristics, issues, and more.

Skydroid C12

The skydroid C12 combines a high-resolution thermal imager and a 2K colour camera[25]. This makes it a dual-lens camera. One lens is an infrared thermal imaging camera that helps in detecting the temperature of the target surface. At the same time, the other one is a regular RGB lens. This helps capture both images in a single flight.

Image 2.6 – Skydroid C12 Dual Camera[26]

Additionally, it has 3-axis stabilization with a built-in gimbal camera. This sensor provides a 25° wide field of view. This and other similar systems weigh around 100 to 600 grams.

Drone-Based Applications:-

- Solar utility inspection
- Powerline monitoring
- Tower inspection
- Structure inspection
- Inspection of pipelines
- Firefighting and rescue operations

2.6 Multispectral Imaging Sensor

Multispectral cameras capture images at different wavelengths of light, such as near infrared, red band, red edge, blue band, and green band, with a precise set of wavelengths for each. Additionally, it does capture a regular RGB image. This image captures all the bands, and RGB happens simultaneously. So, when a particular area is captured in red band, the same area has to be captured by blue, green, red edge, NIR, and RGB lens. This collective data is useful for post-processing. Its key applications are precision agriculture and smart farming.

Micasense Rededge – P

Image 2.7 – Multispectral Camera from Mica Sense[27]

Key Specifications[28]:-

- Spectral Bands - Blue, green, red, red edge, near-IR
- Wavelength - Blue (475 nm), green (560 nm), red (668 nm), red edge (717 nm), NIR (842 nm)
- Multispectral GSD – 7.7 cm per pixel (per band) at 120 m (~400 ft) AGL
- Multispectral field of view - 50° **HFOV**, 38° VFOV
- Weight – 400 grams

Drone-Based Applications:-

- Monitor the evolution of crops
- Track progress of seeding efforts
- Disease identification
- Fertilizer management
- Crop health mapping

2.7 Hyperspectral Imaging Sensor

Hyperspectral imaging is a method in which data is collected and processed, as with the other sensors mentioned above. The difference with hyperspectral imaging is that it collects data and processes information across the electromagnetic spectrum. This way, it obtains the spectrum of each pixel in an image, thereby enabling the identification of different materials and objects by analysing their unique spectral signatures.

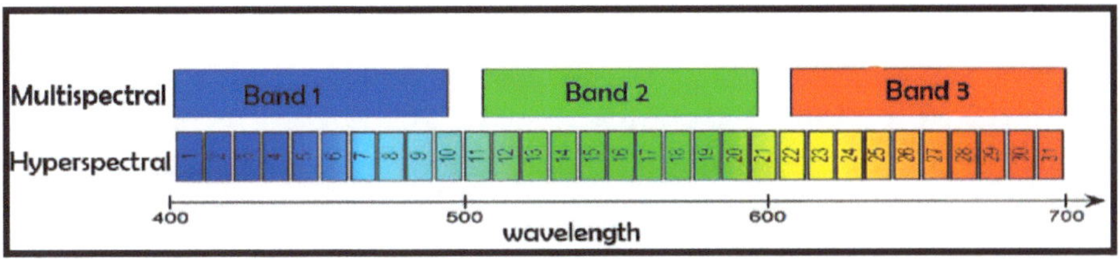

Image 2.8 – Comparison of Multispectral and Hyperspectral Bands[29]

The image above illustrates a difference in the wavelength range captured by multispectral and hyperspectral sensors. Compared to multispectral images, hyperspectral images provide more detailed information, enabling more accurate analysis, identification, and segregation of materials. The key benefit of hyperspectral imaging is its high spatial and spectral resolution, which enables the detailed characterization of materials.

Image 2.9 – Hyperspectral Sensor for Airborne System (Specim)[30]

The image above is of a hyperspectral camera from SPECIM. The key feature of this sensor is that it captures NIR bands ranging from 900nm to 1700nm.

Applications:-

- Water quality monitoring
- Natural disaster monitoring
- Pollution detection
- Agriculture and vegetation
- Oil and mineral exploration
- Climate change analysis

2.8 Agricultural Spraying System

Storage tanks are used as a payload in two different applications. One is an agricultural pesticide-spraying drone for spraying water and pesticides. Another use case is unmanned firefighting, which can spray water at a fire tragedy by going overhead to the spots or where it can reach with horizontal water spraying.

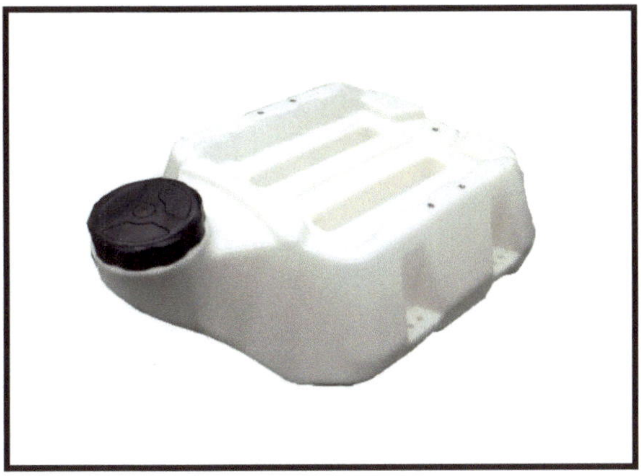

Image 2.10 – Agriculture Drone Water Tank

The image above shows a 10-liter water tank that can be carried as a payload on a drone. However, the spraying tank is not the only payload. Because agricultural drones have many other accessories that also need to be considered part of the payload, viz., pumps, tubes, spraying nozzles, etc.

2.9 Aerial Delivery System

Cool boxes on the ground are mostly used as ice boxes and for some medical storage applications. The use of UAVs has advanced significantly, and so medical and emergency applications should not lag. A normal delivery will require a box to carry the product or medical first aid. However, for special cases such as vaccines and organs, medical cool boxes are required.

Image 2.11 – Cool Box for Human Organ Delivery[31]

The cool box above was designed by Polar Tech Industries to transport a kidney. This life-saving organ-delivery box is equipped with a specially designed wireless biosensor, the Human Organ Monitoring and Quality Assurance Apparatus for Long-distance Travel (Homal). This sensor measures the temperature, barometric pressure, altitude, vibrations, and GPS location. Throughout the experiment, the temperature of the kidney remained stable at 2.5°C. This application of drones is risky but also very important and useful. It is risky when used with multirotor-type drones because, in the event of an incident en route, such as a bird strike or electronic failure, the drone crashes, resulting in damage to the organ, which is very important for the patient waiting for the same. It is important and useful because it takes less time to travel from source to destination, since road traffic is not a hurdle. Additionally, it can reach critical locations in the mountainous region much faster than the road route.

Image 2.12 – Zipline's Delivery Drone[32]

2.10 Defence Payload System

There is a variety of UAVs useful in defense, each designed for different applications, ranging from surveillance to dropping explosives. Unmanned Combat Aerial Vehicles are primarily used for reconnaissance and target attack. These UAVs are equivalent to manned fighter planes and therefore have the capability to return to base. These UAVs carry missiles as a payload along with multiband surveillance sensors. Apart from that, it is always equipped with other sensors and safety-related payloads.

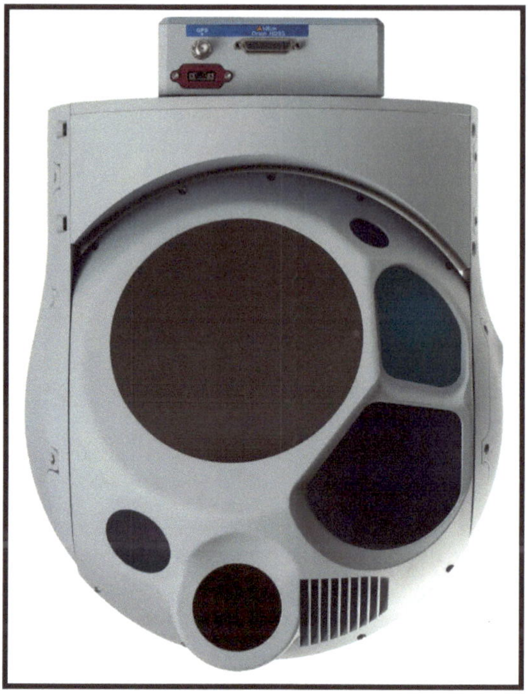

Image 2.13 – EOIR Sensor[33]

There are many drones that carry small or large missiles as payloads, among other payloads, because it is safe and easy for defense applications. Few UAVs are listed here:-

a. USA - General Atomics MQ-9 Reaper
b. Israel - Elbit Hermes 450
c. Turkey - TAI Aksungur
d. India - DRDO Ghatak
e. South Africa - Milkor 380
f. Russia – Kronshtadt Orion
g. China – Hongdu GJ-11

Image 2.14 – Hellfire Missiles[34]

The above image shows a military UAV with two Hellfire missiles in its payload bay.

Points to Remember

- Any designated payload can be added to the drone as required.
- It is to be noted that certain COTS drones come with a camera of the seller's choice. So, it is necessary to analyze the camera type and its specifications, and finalize the purchase if it meets the requirements.
- The selection of the payload is critical for any application, primarily due to weight constraints. It is difficult to select a sensor that best matches the requirements and has the lowest weight.

PART II

DRONE DATA AND MISSION PLANNING

CHAPTER

03

Fundamentals of UAV Data Acquisition and Processing

There are various applications of drones, with almost unlimited possibilities, and new applications are being identified and implemented in no time. All these applications fulfill some or all of the requirements. Some of these use cases are direct, meaning the live data shown or collected can be considered a direct result. At the same time, other applications are called indirect, meaning the data collected using a camera or sensor cannot be considered directly useful for identifying defects and other parameters. This collected data needs to be processed to achieve the desired results.

Overall, the data processing consists of 4 different stages: acquiring data with drones, processing the collected data, analyzing the processed data, and generating reports/deliverables.[35]

3.1 Data Acquisition

Data acquisition is an important factor in data processing and result generation. However, planning for data acquisition is critical, as the final result depends on the collected data, which in turn depends on factors such as the drone and payload used, sensor quality, and data acquisition technique. A customer will have specific and desired requirements. As per the requirement, a particular drone will be selected for the mission. A proper calculation of how this mission will be executed, such that the best outcome is received. When planning the data acquisition procedure, major parameters such as Ground Sampling Distance (GSD) and Ground Control Points (GCPs) are always considered. There may or may not be requirements for Ground Control Points, depending on the project. Projects that require higher mapping accuracy require GCPs, whereas certain inspection-level missions may not require them.

A. GCP

Ground Control Points are reference points used to accurately map large areas and reduce error margins. To get an accurate result, GCPs need to be properly planned and executed. It can be in the form of a structure, i.e., a building, a big rock, etc. GCPs can also be marked at the desired location and at the required distance interval within the targeted area of the map. GCPs can be pre-printed or manually painted. Mostly, data acquisition teams use different-colored spray paint to mark GCPs at targeted locations, such as the roadside, manholes, and rocks. In terms of accuracy, pre-printed GCPs are preferred. Also, they are reusable. In a spray-painted GCP, it is often difficult to find the center point. Additionally, a number must be mentioned beside each GCP. Shown below is the image of a printed GCP being used for a mission. A number is mentioned on the GCP, and the reading is being taken from the exact center point of the GCP.

Image 3.1 – A Printed Ground Control Point[36]

The center point of the GCP is important as it works as the checkpoint, and the levels are measured at this point. Terrestrial surveying equipment is required at each GCP to measure the elevation at that latitude and longitude. This is a must for GPS and PPK-based drones when used for mapping, survey, and volume calculations. For a drone with an RTK system, very few GCPs are required, or none at all, depending on the project. GCP points should be located in an open area with no buildings, trees, or poles obstructing the view of your GCP in drone photography. The clearer the resolution and the visible center-point markings, the more accurate the data will be.[37,38]

B. GSD

GSD stands for Ground Sampling Distance. It is the distance between the centers of two pixels measured on the ground. It is the physical ground distance captured in a single pixel of the image. GSD depends on various factors, such as sensor width, focal length, image width, image height, and altitude at which the drone is flying. All these parameters can be calculated during mission planning to assess whether the mission meets the customer's GSD or data quality requirements. Many times, the client provides the flying altitude. However, in most of the projects, the GSD requirement is provided by the client to the service provider.

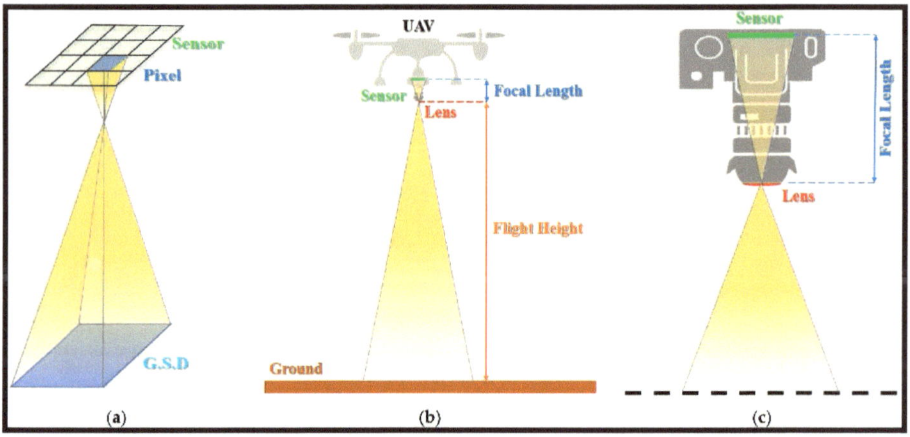

Image 3.2 **(a)** – Pixel Size and its Relation with GSD, **(b)** – Data Acquisition, **(c)** – Details of Main Camera Elements [39]

An example of GSD: a 24 MP survey camera with a fixed focal length of 25mm. The sensor size is 366.6 mm^2 (Sensor width: 23.50mm × sensor height: 15.60mm). The flying altitude is 100 meters above ground level. When a camera captures an image at 24MP, its resolution is 6000 × 4000 pixels (width × height). With all the above required parameters, GSD can be calculated as follows:-

Based on the above calculations, the GSD for images captured at an altitude of 100 meters will be 1.57 centimeters per pixel. The lower the GSD, the higher the spatial resolution, with more detailed images and data accuracy.

The next step after GCP and GSD is planning the mission and data collection.

C. Important Parameters for Mission Planning and Data Acquisition

Proper planning of data collection in the targeted area is crucial because it saves field effort, time, and overall project costs. So, once a service provider receives the targeted plan for the area to be mapped, the team checks the area for any issues and plans the flying mission (one or multiple) as per the provided area in the Ground Control Software. According to the GSD role, the flying altitude is kept the same across all mission parts, as per the desired GSD value.

Another major parameter is flight planning, angle, and grid plans. There are various methods and plans for generating mission plans. This will be covered in the chapter “Application-based mission planning for UAV operations”.

When planning a mission, another important parameter to consider is “Overlap”. Overlap is a parameter useful for mapping and survey applications and must usually be set during mission planning. Overlap is the repetition of a portion that has already been captured in the previous image or will be captured in the next. For example, a person takes a photograph from a height of 2 meters and captures a patch of 1 meter. Again, he shifts a little to capture the image, but covers only half the area of what he already captured in the previous photograph. That is 50% repetition of the area, and can be called the overlap parameter, which is set at 50%. Check the image below for a better understanding. There are two types of overlap. Front/forward overlap and side overlap. Front overlap occurs in the direction the drone moves. Whereas side overlap occurs with the adjacent images captured from along the path. In the image below, the arrow indicates a UAV’s forward path. Each circle represents a photo taken along the flight path, and the square portion is the area captured in each image. The dark area in the rectangular boxes where the front and side overlap is the overlapped area.

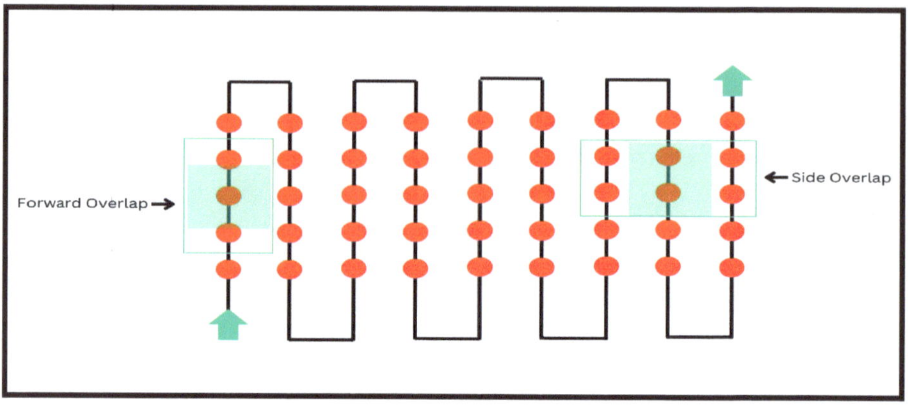

Image 3.3 – Front Overlap and Side Overlap Explanation [40]

Based on the overlap parameters, the grid lines will be presented in the GCS, indicating the desired survey path.

Once all the above planning and settings are done, the drone is ready for take-off and to accomplish the mission.

3.2 Data Processing

Data processing is an essential step that involves multiple methods for extracting valuable information from drone-collected data. The images captured by the drone are raw, so they cannot provide any detail beyond what is visible in the photo. When these images are stitched, they form a single image of the entire area. Desired outputs like 3D point cloud, Ortho mosaic map, 3D elevation model, etc., can only be achieved by processing the collected raw data. To process the data, various software tools are required to execute the process. The features of the software listed below are distinct and therefore have different levels of output capabilities. The data processing software is GUI-based and mostly automatic. If the precise inputs are provided, software can align the images as per the overlap settings and geotagged locations, correct image distortion, etc. A few of these software are listed below with a brief explanation:-

A. DroneDeploy

Drone deploy is a software and application which is helpful for two tasks. Drone deploy has a GCS application. i.e., it is useful for planning a mission and automatic data collection for various different use cases. However, this software has limited compatibility to a few drone models from DJI, Skydio, Freefly, and Wingtra company-manufactured drones. Another useful product they provide is data processing software to process the collected images. Drone deployment is primarily used for image stitching to generate an orthomosaic.[41]

B. Pix4Dmatic

Designed for large-scale, high-precision drone mapping, Pix4Dmatic specializes in processing extensive datasets using robust photogrammetry algorithms, enabling the efficient generation of accurate maps and 3D models.[42]

C. Pix4D Mapper

This versatile photogrammetry software transforms aerial images into detailed 2D maps and 3D models, supporting a wide range of applications including surveying, construction, and agriculture. This software is best for ortho mosaic and DSMs. It is useful when high accuracy is required in the project. It allows customization of users' processing requirements, and only the selected processes will run.[43]

D. QGIS

QGIS is an open-source software useful for data visualization, mapping, planning, contouring, and other geospatial data analysis.[44]

E. AgiSoft Metashape

A comprehensive photogrammetry tool that processes images to produce high-resolution 3D spatial data and models, widely used for GIS, cultural heritage preservation, and visual effects. This software is recommended for agriculture applications.[45]

F. Global Mapper

A powerful GIS software that provides an array of spatial data processing and analysis tools, including terrain analysis, 3D data visualization, and conversion between various file formats.[46]

G. Propeller

Cloud-based software focused on drone data processing for construction and mining, offering tools for generating accurate site maps, tracking progress, and performing volume calculations. Propeller aero is best for construction-related projects. [47]

H. Open Drone Map

An open-source photogrammetry toolkit that processes drone-captured imagery into 3D models, point clouds, and georeferenced maps, promoting accessible and community-driven development.[48]

I. Bentley Context Capture

Enables the creation of high-resolution 3D models from simple photographs, excelling at producing detailed, accurate digital representations of infrastructure and urban environments. This software is recommended for users who want oblique 3D photography. It is recommended for large-scale projects.[49]

J. ESRI Site Scan

A comprehensive drone mapping solution integrated with ArcGIS, providing tools for flight planning, data collection, processing, and analysis to produce accurate geospatial data and models.[50]

K. Sim Active

Sim active specializes in high-end photogrammetry software for processing aerial imagery into precise orthophotos, DSMs, and 3D models, widely used in industries such as surveying, agriculture, and mining.[51]

To process the data, there are two methods. One, the user can download software and perform complete processing in that software. Two, the user can upload the data to the cloud of a particular software platform and process the complete data there. Some of the above software offers cloud-based processing, while others are fully downloadable. The above software may require a subscription fee based on the user's requirements. There are plans available ranging from 1 day to 1 week, to 1 week to a yearly subscription. Data processing is highly automated but may require processing in parts.

3.3 Data Analysis

Data analysis is a critical technology for understanding data patterns. Data analysis is available in each and every domain. This is, in fact, a critical part of any sector or organization. With analysis, it is possible to avoid unwanted situations that could cause disaster. So that improvements and changes can be implemented beforehand. Data analysis is helpful for understanding future market trends, demand and supply, optimizing the supply chain, assessing demand for a new product, and developing marketing strategies, etc.

In drone applications, it is useful for understanding the change pattern. This analysis can only be performed once the data has been fully processed. Post which, the data is analyzed as per the customer's requirement. For example, there is a dry region with some slope and terrain variation. The customer wants to know where the water stream will head in the rain, in which direction, where it will be collected, and whether it will create a dangerous situation, etc. This identification is done through data analysis. Such an activity may include statistical analysis, machine learning, and predictive modeling. So, as per the example, customers can get evidence to understand the problem, which helps them plan the solution properly and work to eliminate any dangerous situation.

Data analysis may not necessarily be an automated system and requires human expertise to understand and apply analytical techniques, check the results, and make decisions based on expert analysis. These results are often presented in visuals and reports. These reports include recommendations for decision-making processes.

3.4 Deliverables

The entire project, or a mission, from data acquisition to processing, depends entirely on the required end results, often called deliverables. These deliverables can be in the form of raw images, reports of the processed data identifying the critical points, dashboard representation, etc. The deliverables are always one of these three: raw data, processed data, and analyzed data. Brief information on the key standard deliverables is provided below; the client may require one or more of them. Additionally, the purpose for which each process is required is specified.

i. Ortho mosaic

Ortho mosaic is a geometrically corrected and scale–accurate aerial image created by stitching the overlapped raw data images collected by the drone. So, to create a single geo-referenced image map (ortho mosaic), multiple overlapping raw images, GPS/PPK/RTK coordinates, and GCP data for the targeted area are to be uploaded to the software as input, which will then be processed.

Image 3.4 – Ortho Mosaic Output of an Area[52]

Purpose:-

- Base mapping for planning roads, plots, and utilities
- Track geometry
- Encroachment detection
- Measurements of carriageway width
- Pipeline corridor monitoring
- Crop boundary and growth stage monitoring
- Manual and AI-based object counting, like trees, infrastructure, etc.
- Panel layout verification and structural damage identification in a solar plant
- Thermal ortho mosaic – hotspot detection and faulty panel identification

ii. 3D Point Cloud (Photogrammetry)

A set of XYZ points representing the shape of objects as per the terrain in 3 dimensions, generated from the images. The creation of a 3D point cloud also requires raw, overlapped images as input data.

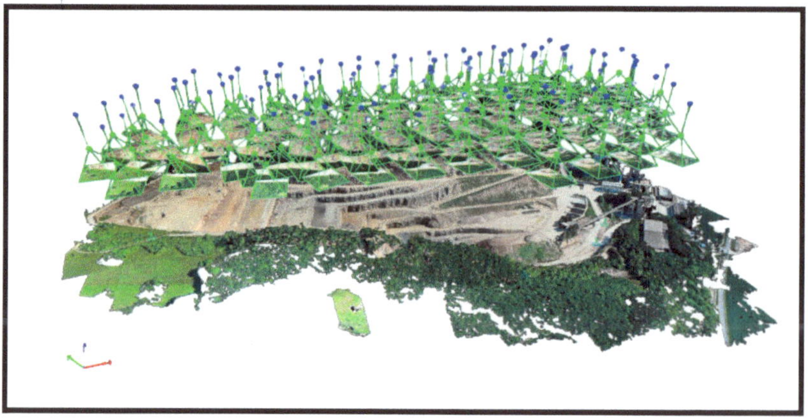

Image 3.5 – 3D Point Cloud of Images[53]

Purpose:-

- Urban 3D modeling
- Void and volumetric estimation
- Shadow studies
- Accurate geometry of structures
- Structural analysis
- Digital asset documentation

iii. LIDAR Point Cloud

A highly accurate 3D point cloud generated directly using the LASER pulses. To generate the LIDAR 3D point cloud raw data from the LIDAR sensor, along with IMU and GNSS data, is required to be used as input.

Image 3.6 – LIDAR Point Cloud[54]

Purpose:-

- Accurate 3D representation of structures and terrains
- Measurement of infrastructures, trees, poles, etc.
- Tunnel deformation
- Clearance analysis
- Line sag calculation
- Track level analysis and cant (superelevation)
- Verticality and deformation analysis

iv. Digital Elevation Model

A general elevation model describing the elevation of the ground surface, excluding man-made structures, vegetation, etc. It represents the elevation of the Earth's surface as a digital image, where each pixel contains the elevation value at its center. DEMs have related topographic terms: DSM and DTM.[55]

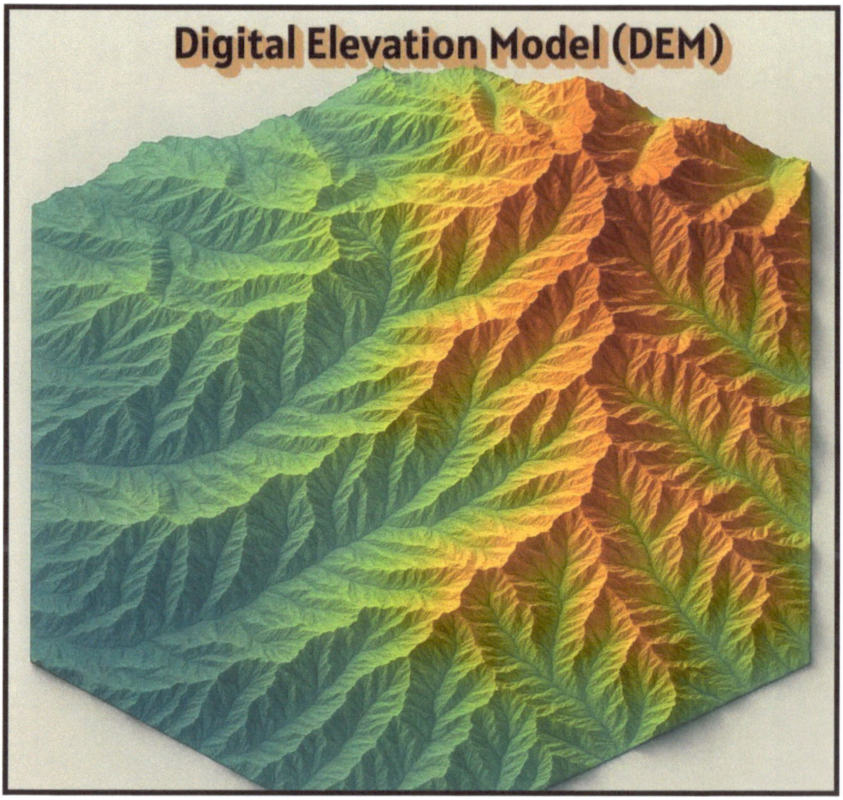

Image 3.7 – Digital Elevation Model[56]

Purpose:-

- Contour analysis
- Slope analysis
- Hydrological modeling
- Watershed analysis

v. Digital Surface Model

DSM is a surface model comprising objects and terrain data, including road surfaces, vegetation, structures, etc. The "surface" refers to the top-most surface of the area. This can be generated using a photogrammetric or LIDAR 3D point cloud.

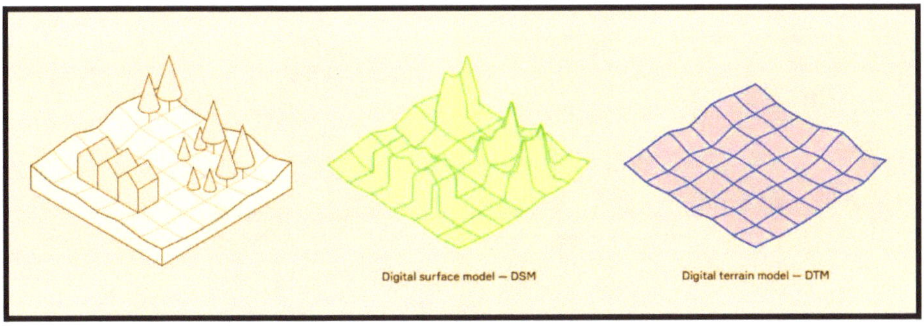

Image 3.8 – Digital Surface Model[57]

Purpose:-

- Height analysis
- Studies of obstruction
- Skyline analysis

vi. Digital Terrain Model

DTM is a terrain model. This is a bare-earth elevation model similar to DEM with no objects - vegetation or man-made structures included in the data. However, this is a more refined version of a DEM. A raw LIDAR point cloud or DSM will be required to generate the bare-earth model/DTM.

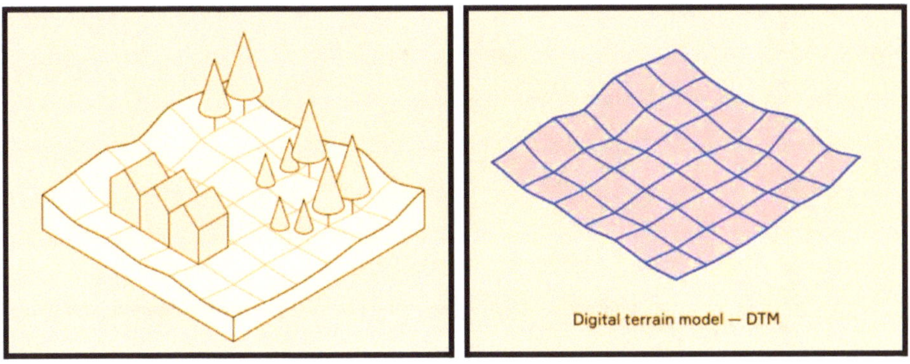

Image 3.9 – Digital Terrain Model[58]

Image 3.10 – Comparison of DSM and DTM[59]

Purpose:-

- Road design
- Drainage design
- Flood analysis
- Road profile
- Cutting and filling data
- Accurate ground elevation reference
- Critical for flow direction and accumulation

vii. Contour Maps

Contour maps are 2D representations of 3D landscapes, comprising lines connecting points of equal elevation.

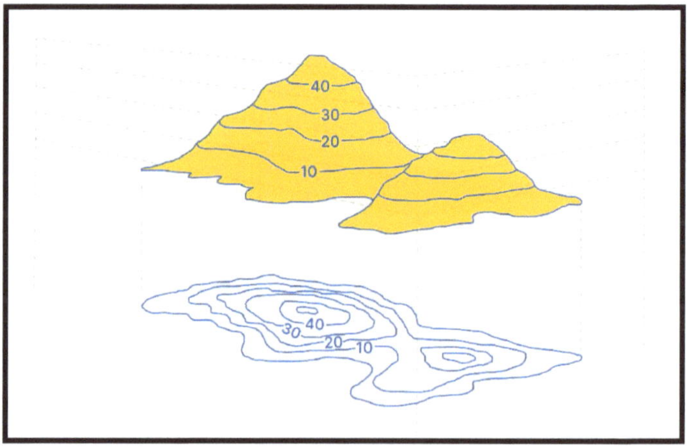

Image 3.11 – Contour Map[60]

Purpose:-

- Terrain interpretation
- Engineering planning
- Slope and elevation variation along the alignment
- Elevation difference visualization

viii. 3D Mesh Model

A surface model created by connecting 3D point clouds into a triangular mesh. The 3D surface with geometry only is received as an output data known as a 3D mesh model.

Purpose:-

- Shape analysis
- Structural modeling
- Surface representation
- Visualization for planners and stakeholders

ix. Textured 3D Model

A 3D mesh that includes a realistic image texture. RGB images are added alongside the 3D mesh file to create a textured 3D model.

Purpose:-

- Realistic visual representation for planning
- Digital preservation and restoration planning

x. Slope and Aspect Maps

The maps were generated to understand the terrain's steepness and direction. DEM or DTM is used to generate the slope and aspect map.

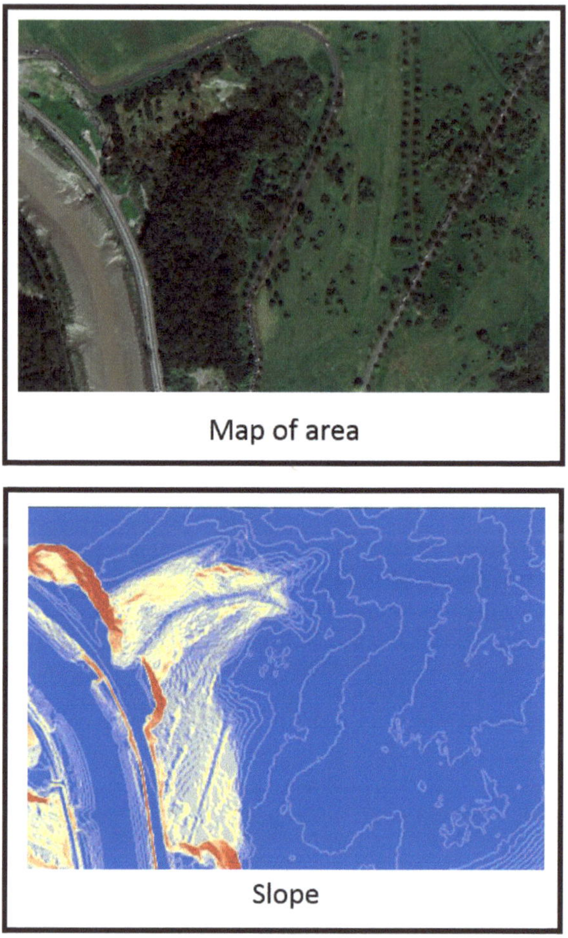

Image 3.12 – Slope Maps[61]

Purpose:-

- Water movement prediction
- Land suitability identification

xi. Thermal Ortho mosaic

An ortho-mosaic map created from geo-referenced thermal images.

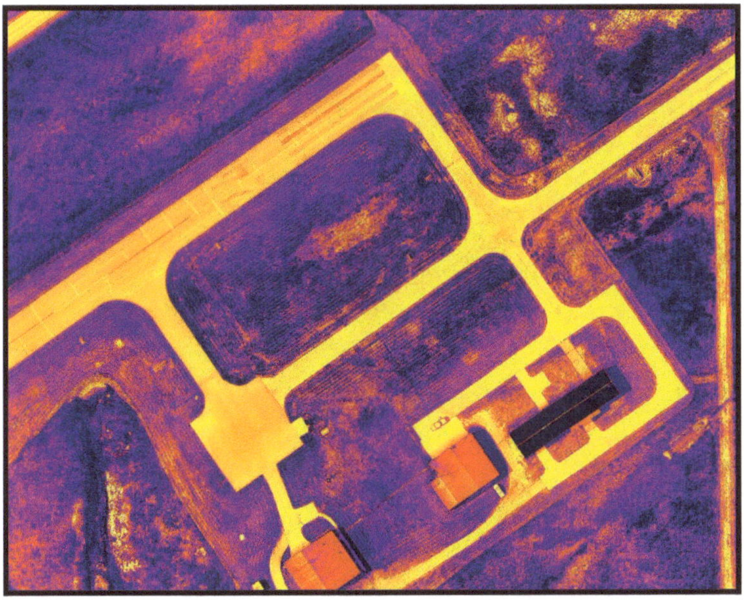

Image 3.13 – Thermal Ortho Mosaic Map[62]

Purpose:-

- Hotspot detection
- Energy audits
- Fault detection
- Faulty panel identification

xii. High-resolution RGB images

Raw images captured by the RGB camera.

Purpose:-

- Inspection of components like bolts, insulators, etc.
- Physical damage identification
- Surface crack detection
- Corrosion detection
- Other visual inspection
- Documentation

xiii. Thermal Images

Thermal images are raw images captured from thermal camera sensors.

Purpose:-

- Fault identification
- Hotspot detection
- Overload detection
- Overheating identification
- Internal fault identification

xiv. Thermal Data

The temperature matrices chart with numerical temperature values extracted from thermal images.

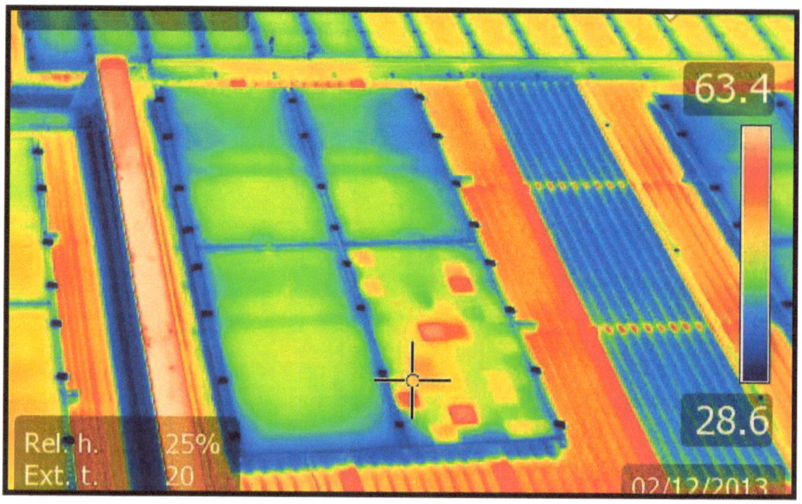

Image 3.14 – Thermal Data[63]

Purpose:-

- Temperature anomalies indicating leaks
- Quantitative analysis

xv. Anomaly Maps

Anomaly maps highlight abnormal patterns. These maps are generated using RGB data and thermal data.

Purpose:-

- Early fault detection
- Maintenance decision support

xvi. Vegetation Index Maps

Vegetation index maps are generated from multispectral data, providing details about vegetation condition. These are index maps – NDVI, SAVI, GNDVI, etc.

Purpose:-

- Stress and variability analysis
- Crop analysis

xvii. NDVI Map

A specific vegetation index map showing plant health was generated with red band and near infrared band images.

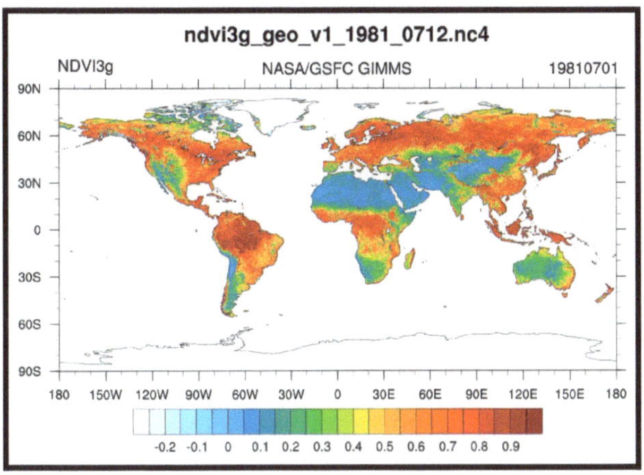

Image 3.15 – NDVI Map[64]

Purpose:-

- Chlorophyll and biomass estimation
- Crop stress monitoring

xviii. Zonal Statistics

Zonal statistics are values providing statistical summaries within the defined area. It is generated using an NDVI raster or other raster maps along with its field boundary data. The output results are the mean value, minimum/maximum values per zone.

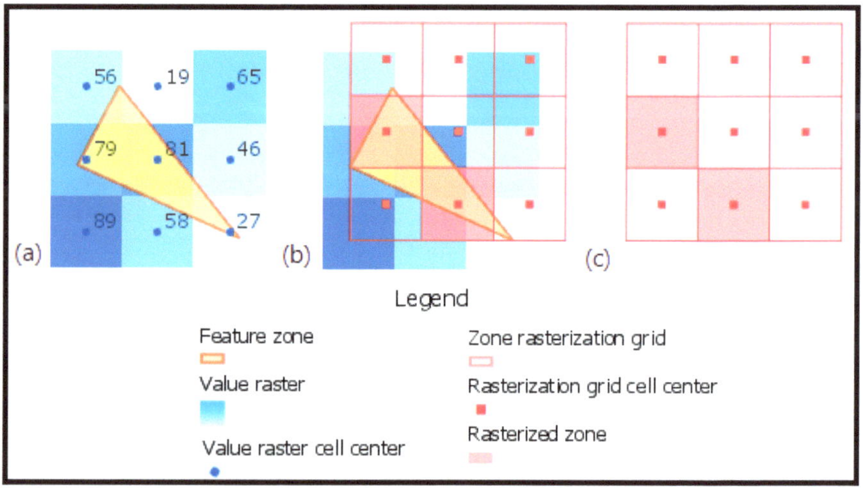

Image 3.16 – Zonal Statistics[65]

Purpose:-

- Field – wise crop health comparison
- Farm level decision

xix. Cut-Fill Reports

Cut-fill reports are calculations of earthwork volume that can be generated using DTM and DSM.

Purpose:-

- Earthwork estimation
- Road widening
- Construction estimation

xx. Inspection and Defect Reports

Inspection reports are the consolidated findings from the image-based analysis done manually or with AI. Whereas the faults identified are mentioned in the analytical reports. Inspection reports comprise RGB and thermal images, along with measurements, to produce a structured report. To generate the defect reports, RGB and thermal images, as well as inspection results, are required.

Purpose:-

- Compliance and safety
- Maintenance prioritization
- Safety planning
- Facility management
- Risk mitigation planning
- Structural safety evaluation
- Asset health assessment
- Maintenance planning and shutdown optimization

Result

The results required for all survey, mapping, inspection, etc., projects are mostly in the form of a report. Additionally, a few customers require a dashboard representation to display results to their customers.

3.5 Satellite Vs. Drone

The initial discussion about drones and data processing is complete. However, there is always a question. Why drones, when we can collect data using satellites? There are many differences between the data collected by a satellite and a drone.

A. Scale and Resolution

Drones provide operational flexibility, as they can only capture the required area, whereas satellites capture the target area, including additional areas. This happens due to the altitude at which images are captured. Controlling a drone's altitude is easy. The image below shows an island captured using a satellite and a drone. In comparison, the satellite is collecting more data than is required. Whereas the data collected by the drone is limited to the particular selected area.

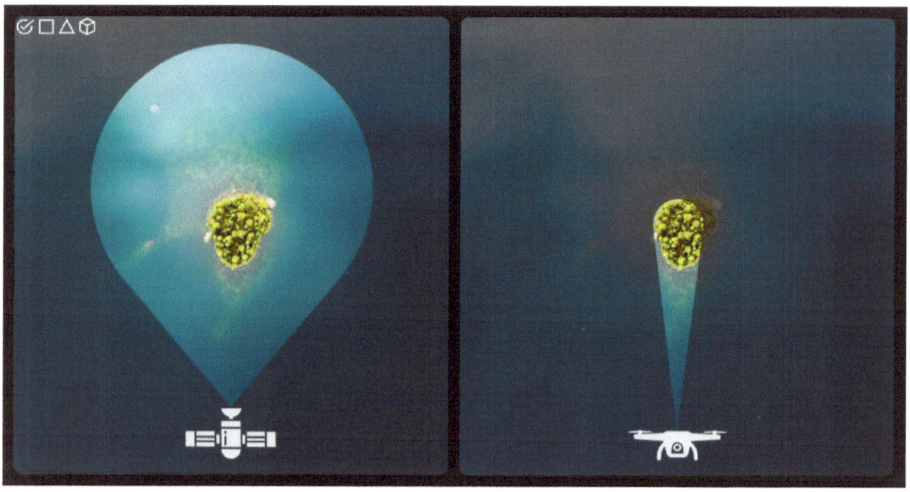

Image 3.17 – Satellite vs. Drone Data Collection[66]

B. Sensors

Satellites and drones both use a variety of sensors to capture images. Satellite sensors capture data via electromagnetic (EM) radiation, and to record it, synthetic aperture radars and multispectral sensors are required. Similarly, drones can use different cameras/sensors, such as RGB, thermal, LIDAR, multispectral, EOIR, depending on the capture requirements.

C. Ground Sampling Distance

Drones have high GSD ranging from cm to mm/pixel, and very high GSD when it calculates less than 1mm/pixel. Whereas in satellites, its GSD is considered high when it ranges from 1 meter to 5 meters per pixel. This affects the details captured in an image.

D. Recording Duration

The satellite captures data throughout the day and night. However, it takes days for a satellite to revisit the same location. Drones capture data based on their range and endurance. It is flexible to recapture the incorrectly captured data within the very next hour.

E. Sensors

Satellites carry multiple sensors at a time, whereas drones carry sensors as needed. The sensors that the satellite carries are optical sensors, thermal infrared sensors, radiometers, synthetic aperture radar, etc. Whereas drones carry visible-spectrum cameras, LIDAR, hyperspectral and infrared sensors, GNSS, and IMU.

F. Colors

Drones and satellites capture images in different colors. The common capture styles between the two are visible spectrum (RGB), thermal imaging, hyperspectral, and multispectral imaging, among others.

Due to their affordability, data quality, and precision, UAVs are preferred over satellites for certain data acquisition use cases.

Points to Remember:-

- With lower GSD, accuracy will be high. However, the flying altitude will need to be kept lower. Which means the flight time will be longer, resulting in a greater number of images.
- With higher GSD, accuracy will be low. However, the flying altitude will be high. Which means the total flight time will be shorter, and the data collected will be less than with high-accuracy collection.
- The workflow and technique will change slightly when the data is collected using RTK and PPK systems.

CHAPTER 04

Application-Based Mission Planning for UAV Operations

UAVs are designed and developed with the most advanced electronics and sensor systems, enabling them to fly autonomously. Autonomous operations are software-controlled smart operations in which there is minimal or no need for humans to control the vehicle. To carry out such an autonomous operation, a decision needs to be made regarding target requirements and to plan the operation. UAV operations can be planned, executed, controlled, and monitored using a ground control system. Setting up the path, altitude, flying speed, where to stop, where to land, where to take off, flying at different altitudes, dropping a payload, etc., are necessary parts of mission planning, which can be done using a GCS. Once proper planning is carried out on the software and applied to the vehicle, drones will follow these paths and perform the required operations. In order to understand "How to plan a mission", it is necessary to understand what ground control station software is and how it works.

4.1 Ground Control Station

GCS stands for Ground Control Station, which can be any device, such as a remote controller, smartphone, or laptop, with the corresponding ground control software active on the device. A GCS communicates with the drone wirelessly via a telemetry system that needs to be connected to the GCS at one end and to the drone's autopilot at the other. Once the correct connection is established, it will start showing the vehicle's updated parameters continuously. In this GCS software, there are options to configure the drone with required inputs, such as the drone type and battery specifications. Once the required settings are in place, calibration settings are required. An autopilot system should be mounted on the drone so that the autopilot's forward arrow aligns with the front of the vehicle. However, this is not always possible, so there are facilities to change the orientation of the flight controller board and the offset mount. This is allowed in many advanced flight controller systems where the offset angle is to be set so that it can nullify the offset, and controls can function properly.

Similar settings, planning, execution, and monitoring vary with the changes in the Autopilot system development. So, different flight controllers will have different settings and options available for carrying out the operations. Indeed, the major parameters remain the same. These parameters will be in miscellaneous terms and are explained collectively in a common format. It is possible that any of the below-explained parameter settings or configurations may not be available in the ground station software that you are using or are planning to use. It is also recommended that, for a drone, the types of missions to be carried out be decided before purchasing a flight controller board/autopilot system, so that no desired operational requirements are missed. The illustrations shown here for explanation may depict a particular type of software system, but the main concept behind all GCS software remains the same.

There are hundreds of parameters updated; however, not all can be listed. Some of the major parameters are:-

Satellite connections, telemetry signal strength, flying altitude, moments of the vehicle, flying speed, distance to the home location, current location of the drone on a map, battery usage, errors, etc. Ground Control Station can be of different sizes. For a military application, it can be in a pilot cabin with a few large screens and two or more control systems to manage changes and control the vehicle. Such ground control stations are connected and operated through satellite communication systems. Whereas for commercial use cases, GCSs are nowadays mostly compact and available as handheld devices known as remote controllers. These remote controllers include a smartphone-like Android device built in. This device allows users to download the compatible GCS application, configure all settings, manage flight operations, etc. The radio communication devices are easy to carry in the field but have a limited operating range of 5 to 20 km. With GCS built into the remote controller, a separate telemetry system is not required unless the user wants to extend the remote controller's range further.

4.2 Head-up Display

A GCS includes systems such as a dashboard, flight parameter settings, and mission planning. A Head Up Display (HUD) is one such part of the GCS system that displays several parameters important to view and note while the UAV is flying. To illustrate the GCS, a software mission planner is used as an example. The overall main screen of the mission planner application allows us to connect or disconnect the GCS from the drone. This connection is possible via a telemetry system or a remote controller connected to the autopilot. Once the drone and the GCS are connected, the current status of the drone and flight controller parameters are displayed on the GCS's main screen. The image below shows the main screen/home of a GCS.

The home screen of the GCS covers the following things:

a. On the top, it provides different tabs/pages to proceed with settings. These include flight plan, parameter settings, simulations, etc.

b. The top right is the connection button to connect with the craft

c. On the left, there is a HUD that shows a complete aircraft operation parameter visual. The HUD will move as the craft's angle changes, and angles are displayed for reference. Other parameters displayed include yaw angle, speed, battery voltage, telemetry connection range, errors, drone level, altitude, etc.

d. Below the HUD, there are various buttons to check and perform different tasks, such as "QUICK", which shows important and selected parameter readings in big and bold. "ACTION" is for quickly taking action, such as pressing an RTL button during an emergency.

e. The map is to show the exact location of the aircraft on the map

f. Below the map, there is a communication satellite count shown

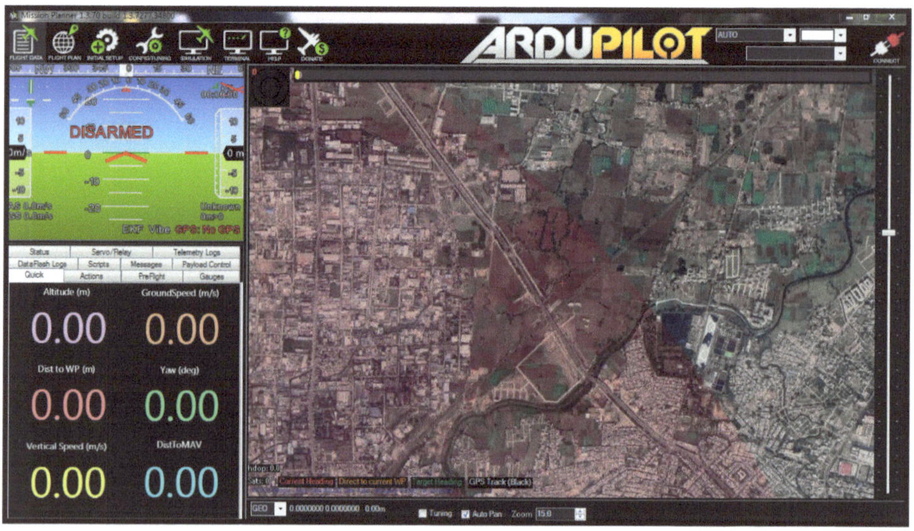

Image 4.1 – Mission Planner Application – Flight Data Screen

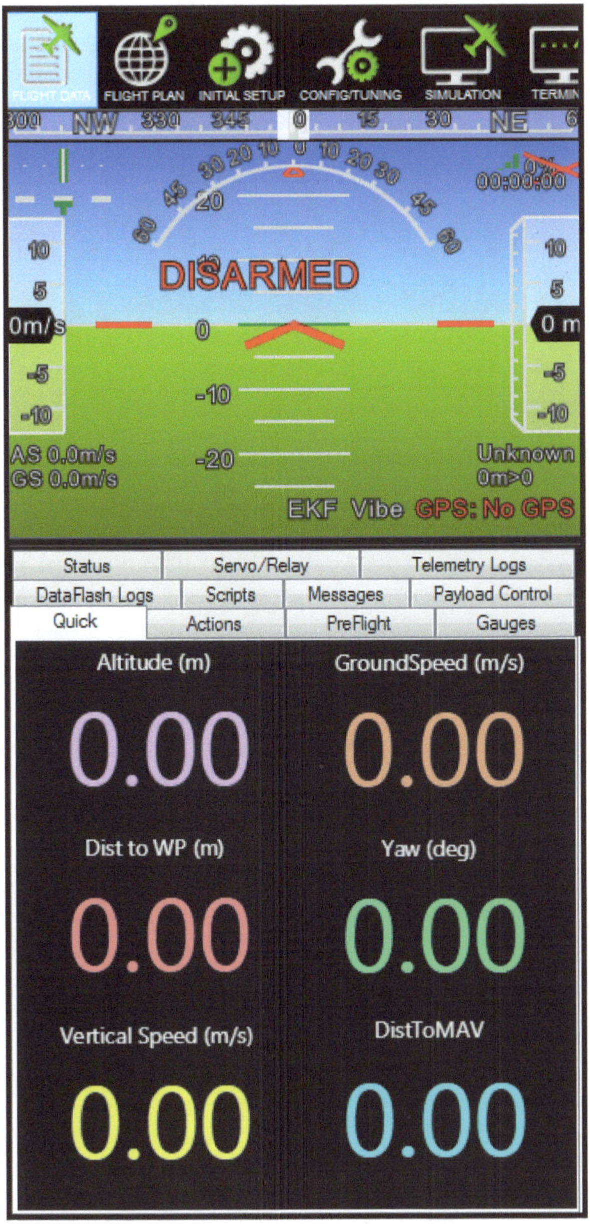

Image 4.2 – HUD and a Quick Screen that Provides Information During the Flight

Throughout the mission, it is easy to identify parameters such as distance from home, flying altitude, speed, climbing speed, vertical speed, vehicle location, remaining battery, and more. The image above shows parameters that can be monitored during the mission on the flight data screen.

4.3 Mission Planning

A mission planning is a process that allows you to create a flight plan so that the drone can fly autonomously on the provided trajectory. Missions can be planned in different formats, like waypoints, rally points, grid missions, and terrain-following. These missions can be planned from the office and saved for future execution. This can save much time and effort during field executions.

It is important to check and set the home point when not on the ground. Basically, when the drone gets armed, the home point is set. With the home point set, the vehicle can return to the home location when RTL is executed.

A. Waypoints Mission

In a multi-waypoint mission, the points are added to the path where the user requires the vehicle to fly. Multiple waypoints can be added by just clicking on the desired locations on the map. Altitude can be set differently for each waypoint.

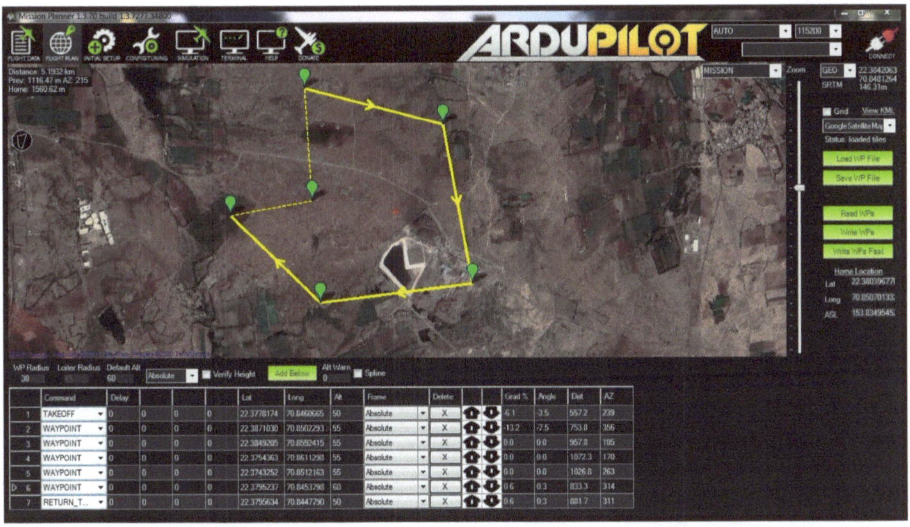

Image 4.3 – Example of Waypoint Adding Up in the Mission Planner

In a waypoint mission, as shown in the above image, five waypoints are added; more waypoints can be added, and, at the bottom, the list of waypoints is provided, from where the order of the points can be changed. Also, other parameters can be selected directly for takeoff, landing, or return-to-launch commands, so they can be executed autonomously.

Such waypoint missions are mostly useful for applications such as training and testing, anti-drone testing, and payload delivery trials. This can be called mission planning for beginners, so they can understand how it works in a small area.

B. Linear Mission

The linear missions are almost straight, with a narrow but long area. As a result, the drones are mostly required to move in a single line, unlike in grid missions, where they follow a zigzag pattern within a particular area. The major applications that require a drone to perform linear missions include railway corridor inspection, power line inspection, bridge inspection, state and national highways, metro rail corridors, oil and gas pipeline inspection, etc.

C. Auto Grid Mission

Grid missions are a type of linear mission in which a drone flies from point A to point B along a straight path. When multiple linear paths are connected for the survey, it becomes a grid mission. These grids are similar to the gaps between two crops in a farm.

While planning an auto grid mission, the grid area of interest for drone flight is provided, and the vehicle executes the mission in a back-and-forth "lawnmower" pattern over the selected area. This grid can be created by selecting the boundary points of an area, and it is automatically generated according to the default settings. After this, other flight parameters need to be set according to the requirements. These parameters include the camera make and model (for best compatibility), the drone's flying altitude, flying speed, grid angle, and image-capture settings, among others. Also, such grid missions can be planned by uploading a KML and then setting parameters for flight and photogrammetry.

Grid missions are useful for applications such as land surveying and mapping, photography of the farm and the solar farm for inspection reference, agricultural spraying, crop monitoring, etc.

Image 4.4 – Auto Grid Mission in the Mission Planner

D. Double Grid Mission

Double-grid missions are similar to grid missions and are also known as cross-grid missions. This is an add-on mission that continues a grid mission: after the regular grid mission is completed, the drone starts flying in the cross grids developed perpendicular to the grid mission.

For example, if the grid mission is planned from South to North, the perpendicular grid will be formed from East to West.

To plan a grid mission, no specific step is required, except that the complete grid mission procedure must be followed, including all flight settings. A toggle button that asks about mission needs must be selected to enable the auto grid mission. The planned mission may not finish in a single flight. Under such conditions, the drones can be returned to their home location, or they will autonomously return home after calculating the remaining battery life to complete the mission, and can resume from the same grid line where they left off in the last captured image after the battery is replaced.

Cross-grid missions are mostly useful for high-precision land mapping and survey. These missions can be performed by all the types of aerial vehicles - VTOL, fixed wing, and multirotor systems with a defined autopilot system.

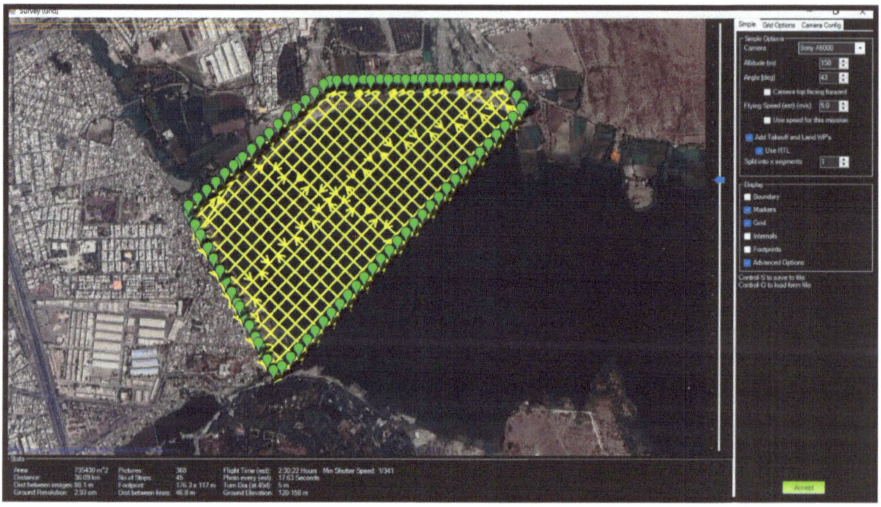

Image 4.5 – A Double Grid Planned Mission Over a Portion of a Lake

E. Circular – Tower

Circular inspections are typical inspections in which the target structures are designed so they cannot be surveyed or inspected regularly. These include telecommunication towers, high-extension power line lattice tower structures, overhead water tanks, chimneys, cooling towers, bridges and pillars, etc. Whatever their shape, these assets must be inspected for defects and undergo periodic inspections. There are very few ground control station software packages that support such an inspection in autonomous mode. One such application is Hammer Mission.[67]

In these types of missions, it is important to be extremely calculative and careful, whether the drone is flown in manual or autonomous flying mode. The required calculations are: tower height, diameter of the largest area, how far the drone will fly from the tower, camera angle requirements, etc. Whether opting for an autonomous or manual mission, it can be flown in two ways to capture the assets, as per the final requirements. Omnidirectional inspections are carried out only with multirotor drone systems.

F. Agriculture farm

Drones are mainly used on farms for two primary applications: monitoring crops and spraying pesticides. Both applications follow grid missions for crop monitoring and pesticide spraying, and, depending on the mission requirements for inspection or pesticide spraying, their parameter settings will vary.

The detailed methodology for planning and executing this mission is explained in the upcoming chapter on drones in agriculture.

Points to Remember:-

- Ground Control Software is mostly GUI-based, and it is recommended to have extensive hands-on experience with the entire system, mainly because, during operations and emergencies, finding the correct button to press can be difficult.
- Missions must be planned as efficiently as possible to minimize the team's location switching.
- Quick decision-making is required of a GCS operator, and the operator must remain ready for it.

PART III

INDUSTRIAL AND COMMERCIAL APPLICATIONS OF UAV TECHNOLOGY

CHAPTER

05

UAV Applications in Survey and Mapping

Surveying and mapping are technologies used to measure and represent the physical properties and boundaries of the Earth's surface. For the development or upgradation of any infrastructure, calculations and planning are required. To better understand the region and plan the management and maintenance of land and infrastructure, foundational activities, i.e., mapping and surveying, are required. Before proceeding ahead towards its use and importance, the concern is to understand why they actually matter.

Mapping is the process of using spatial data to generate maps that include a variety of features, such as boundaries, land, and infrastructure. Surveying provides the numerical and geographical basis for maps. It includes high-accuracy measurement of distance, angle, elevations, and positions.

Overall, mapping and surveys provide us with accurate maps, enabling organizations to plan layouts for various infrastructure projects involving roads, railways, bridges, and building design. Additionally, this provides us with legal clarification on land area and reduces risk with safe zone identification.

A list of a few important mapping and survey applications is given below:-

1. Road widening
2. Town planning and infrastructure development
3. Area elevation calculation for various applications
4. 3D model of the area
5. Water flow analysis
6. Volumetric analysis
7. Object Counting

If these applications are segregated by use case, the number of drone applications in mapping and surveying will be large. Some of the important applications are covered here.

5.1 Road Widening

Road widening is the process of increasing the width of roads, including urban and rural roads and highways, as part of a traffic decongestion drive. The widening helps commuters by saving them time while traveling, as vehicles have enough space to move. However, the same widening becomes a problem for nearby residents, hoteliers, offices, organizations, building infrastructure, farms, etc. The major concerns for a planner include shifting electric poles, managing high-voltage lines, cutting down trees, shifting billboards, cutting down hills, providing drainage facilities, etc.

The use of UAVs in road-widening applications for mapping and surveying can play an important role in saving time and effort, and in capturing sufficient data from every accessible location, including those that are not, by flying over land and even crossing small water bodies, farms, and infrastructure.

5.1.1 Problems and Challenges

The existing problem with mapping and survey is not one. There are many. The challenge starts with manually surveying hundreds of square kilometers. The next line of the problem starts after that. Below is the list of the problems pertaining to traditional mapping and survey methodology.

- Unusual terrain
- Area can be slippery, dangerous
- Not everywhere can humans reach
- Permissions to enter the farm field
- Not everywhere do vehicles have access
- Manpower struggling for the entire day under the sun/cold weather
- Long duration of working and walking in the field
- Time consumption is higher
- Physical counting of infrastructure, trees, poles, ad boards, etc., is much more difficult

5.1.2 Drone as a Solution

There are machines available that can do the same work with less effort, faster, and more efficiently. The expectations are not that humans should stop working. Rather, this is the time to focus on smart work using machines.

To understand the drone-based solution for highway survey or road-widening applications, consider a state or national highway connecting more than two or three cities, with a length of around 300km. Now this is where the difficulty starts for the manual survey. A surveyor needs to keep moving from one location to another to take readings, but in a very small patch. The time required to collect the manual data can be calculated in weeks, whereas the time required to collect the same data using UAVs can be calculated in days. Chances are high that data for a

300km-long national highway can be collected within 10 to 15 days, at an average of 20 to 30km of data capture per day, using UAVs.

Another point to consider is the manpower required on the field. This is important because the faster the process, the less time is required to complete the entire work. So, more output is required in a single day. In such a case, by the traditional survey method, more teams need to be deployed on the field. Whereas, with a mapping UAV, two to three people in a team are enough to cover a few kilometers in a single day. There are many benefits of deploying UAVs in the mapping field, as listed here -

- UAVs can fly anywhere. There are no issues like unusual terrain, steep locations, or entering dangerous areas
- Can be flown from any plain patch beside a highway
- Team member feels safe
- Lower time consumption for data collection

In today's competitive landscape, saving time and achieving better accuracy are the two most important factors to consider when evaluating a project.

Image 5.1 – Drones in Road Widening Survey[68]

5.1.3 UAV System Payload and Flight Operation

As a payload, a high-resolution RGB camera with 24 MP or more is enough. Higher resolution is important when the camera's capabilities are considered for GSD calculations. (GSD has already been covered in the previous chapter pertaining to data processing). This RGB camera can be mounted on an appropriate gimbal for better stability. This is not mandatory but preferred for clearer, better images.

Another payload that can be used for highway surveys and road widening is a LIDAR sensor. LIDARs are costly and require the user to spend more on the project. However, LIDAR provides higher-accuracy results than an RGB camera.

For the selection of UAVs to deploy with an RGB camera, both multirotor and fixed-wing UAVs are preferred and used for highway surveys. In fact, for a long-distance mission, especially when the target area is a linear path, drones with longer endurance are the best choice, and these are always fixed-wing vehicles. However, fixed-wing UAVs need a small, straight, and levelled patch for take-off and landing.

5.1.4 Data Capturing and Processing

Long-distance missions are mostly BVLOS and are therefore carried out in autonomous flight mode. The linear path is the most commonly used mission planning method, as it involves flying along a linear path, like highways. The same can be planned for both fixed-wing and multirotor UAVs.

The mission data collected from the RGB camera are always in the image format, whereas the data collected from the LIDAR survey is in LAS (LASER) format, comprising 3D point clouds. The collected image data is raw, with images geo-tagged. Geo-tagging is an important factor in identifying the exact location of the area in the image. The captured images must be checked for their quality.

Key deliverables and their objective:-

Output	Purpose
Ortho mosaic	– Measurement of carriageway width – Visual identification of utilities – roads, trees, etc. – Object counting
LIDAR Point Cloud	– Accurate 3D representation of structure and terrain – Height measurement of utilities – Clearance analysis
DSM	– Obstruction identification – Height analysis
DTM	– Road profile – Cutting – filling – Drainage design
Contour Maps	– Slope and elevation variation along alignment

Table 5.1 – Road Widening Outputs

5.2 Town Planning and Infrastructure Development

The usage of land, public spaces, housing societies, transportation routes, etc., is organized by the process of town planning. With the help of which, the development of any particular village/city can be carried out systematically, resulting in the reduction of congestion in residential areas and slums. Town planning divides and allocates land for residential, commercial, and industrial purposes. This definitely includes the preservation of forest and agricultural land.

The systematic planning of underground routes for water supply lines, drainage, electricity, and fiber-optic cables may result in cheaper infrastructure than unplanned development in the town. Proper town planning is also beneficial in reducing disease risk by providing fire safety and natural drainage channels, and reducing waterlogging. Well-planned towns attract investments, improve logistics, and boost property values.

Town planning is the basic requirement and foundation for making our society habitable, sustainable, and economically viable settlements.

5.2.1 Problems and Challenges

Updated maps, 3D building models, and terrain models are among the data required to be collected for town planning and its implementation. However, updating maps manually with ground surveys takes months. The details and up-to-date data of 3D building and terrain models are limited. The data can be collected through manned aerial surveys, which are very costly and therefore considered an out-of-budget solution for municipalities.

Manually surveying on foot for thousands of hours in narrow lanes, hazardous sites, disaster zones, and other challenging environments is much more difficult and carries significant risk to life, especially in rough terrain.

5.2.2 Drone as a Solution

Unmanned aerial vehicles reduce risk and human presence on site, which is the first benefit, and, in addition, drones are the cheapest and most effective solution to the above-mentioned problems. UAVs are rapid and can produce ortho mosaics, 3D point clouds, DSMs/DTMs, etc., within hours to days, depending on the area covered.

The data collected and processed are accurate to the centimeter level compared to the manual data, which often has errors. Additionally, there are no road-closure issues for UAV-based data collection. With older and new data collection, there can be a simple comparison and identification of images, focusing on illegal construction, encroachment, etc.

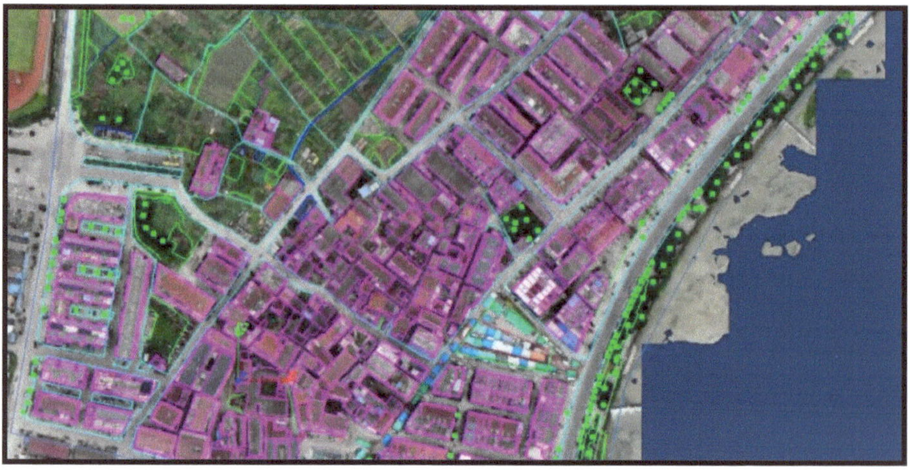

Image 5.2 – Drones in Urban Planning[69]

5.2.3 UAV System Payload and Flight Operation

UAV used for mapping purposes of town planning applications mostly uses a high-resolution RGB camera. These cameras can range from 24 MP to 36 MP resolution. Gimbals, to support the camera, are not mandatory but recommended to avoid the risk of a blurred image. For short- or small-range applications, multirotor systems are the preferred vehicle type. However, for an entire village, town, or city survey, a fixed-wing or VTOL UAV is recommended.

5.2.4 Data Capturing and Processing

In the case of village/town planning, the area is not always in a long, straight path. The target areas are mostly awkwardly shaped, where autonomous mission planning is preferred. Even a large area can be planned in one go, and flight operations can be implemented accordingly. These planned missions have navigation points and routes to be followed autonomously by the UAVs.

The data collected will be in image form. These images are overlapped on the side and front, which helps capture data with higher accuracy and reduces the likelihood of missing data during collection. Once these images are collected, they will be stitched, and further processing to obtain a 3D point cloud, DTM/DSM, ortho mosaic, etc., can be carried out using various software.

Key Deliverables and their objectives:-

Output	**Purpose**
Ortho mosaic	– Base map of planning road, plots and utilities
DSM	– Building height and skyline analysis
3D Point Cloud	– Urban 3D modeling – Shadow studies
3D Mesh/Textured Model	– Visual representation for planners

Table 5.2 – Town Planning and Infrastructure Development Outputs

5.3 Area Elevation Calculation

Area elevation calculation is a spatial attribute that helps convert a flat map into realistic terrain. It is not a separate application. In fact, in all of the above applications, area elevation is required in the result for some reason or another. However, there is a chance that a user/customer may require only the elevation values for the targeted area. Under such a condition, the data-capturing procedure remains the same. However, there is a limited process you will need to undergo, rather than the full-fledged data processing required in the above applications. Certain data processing procedures will not be required to be followed, as this is a simple and straightforward requirement; hence, it will take the least time to process the data compared to other mapping applications. The procedure of data collection using a drone remains the same. Both multirotor and fixed-wing UAVs can capture data for elevation calculations. However, it depends on the total area that needs to be captured.

Key Deliverables and their objectives:-

Output	Purpose
DTM	– Accurate ground elevation reference
DEM	– Slope analysis
Contour Maps	– Elevation difference visualization
Cut – Fill Reports	– Earthwork estimation

Table 5.3 – Area Elevation Calculation Outputs

5.4 3D Model of the Area

The 3D model of an area is a type of mapping, and the easiest way of creating a 3D model of any target area, which can be a town, village, city, or even a society. Any particular area contains a variety of structures, such as trees, apartments, high-rise buildings, a small shop, a pole, a parking lot, and a temple.

5.4.1 Problems and Challenges

Mapping the flat land may take a long time, manually, but it is still easy (not when the terrain is critical). However, when it comes to physical vertical structures, it becomes very difficult to create a 3D model of an area using manual calculations. The biggest issue may not be time; accuracy is.

5.4.2 Drone as a Solution

UAVs for mapping make these tasks faster and more accurate by using a data-collection method similar to that used in most other mapping applications.

Image 5.3 – 3D Model of an Area[70]

5.4.3 UAV System Payload and Flight Operation

Apart from a regular camera for mapping, there is also a sensor that speeds up data collection and processing compared to the standard drone mapping method. A 5-in-1 camera is a sensor that can be attached to the drone instead of a regular mapping camera. This single unit has five cameras mounted at angles in all four directions, plus a camera facing downward (Nadir view). So, all four directions are captured at the required angle as well as the top view. This reduces the total number of images captured across multiple degrees, thereby reducing the total duration of data acquisition. The overall process is very fast, especially when the area is large, such as capturing an entire city. The image below shows a camera sensor unit with five cameras attached. These advanced sensors increase efficiency and reduce the project's overall timeline.

Image 5.4 – An Oblique Camera[71]

The flight operation remains the same for other mapping applications. The plans can be created as per the requirements of the mission – linear or grid mission.

5.4.4 Data Capturing and Processing

The 3D model of an area is easiest with an oblique camera when combined with the grid mission. Whereas it is also possible to create a 3D model of a tall individual structure, such as towers, without a grid mission using other methods.

One such method for vertical structure is to use the available mission-planning applications directly, such as Hammer missions. Hammer is an application used for a variety of missions. However, vertical structure mapping is an extraordinary provision in their application. The software calculates the flight path around the tower based on the input data provided by the pilot, like altitude, horizontal distance, etc.

It captures the data in a circular manner and moves vertically up and down based on the structure.

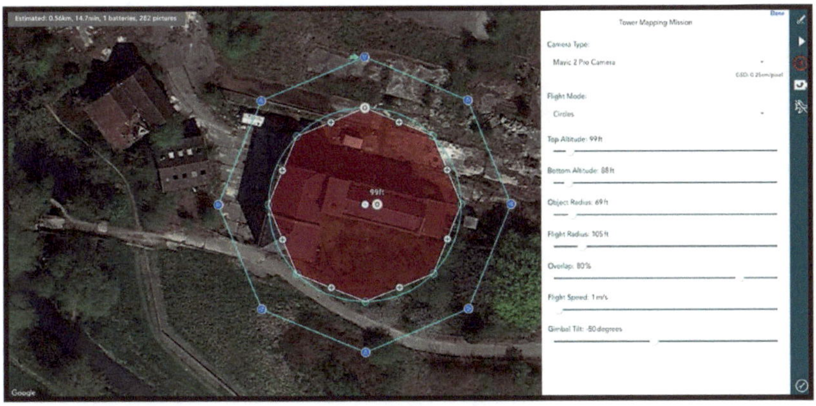

Image 5.5 – Tower Mapping in Hammer Missions[72]

Key Deliverables and their objectives:-

Output	Purpose
3D Point Cloud	– Accurate structural geometry
3D Mesh Model	– Surface representation
3D Textured Model	– Realistic visualization for planning and representation

Table 5.4 – 3D Model Outputs

Points to Remember

- Always remember and pay attention to GSD calculation and to clear, accurate GCP marking in any mapping and surveying applications.
- When planning missions for mapping and survey applications, consider using PPK/RTK drones.
- Consider the range and endurance parameters for planning the missions. It is always better to plan missions in parts, based on the drone's range and endurance, rather than on a single flight's performance.

CHAPTER

06

UAV-Based Monitoring and Surveillance Systems

UAV technology is evolving rapidly, expanding the use cases for drones across sectors, from manual flight to autonomous pattern missions. UAVs are capable of playing critical roles in upgraded situational awareness and proactive decision-making. These vehicles are equipped with advanced cameras and sensors, along with long-range data transmission capabilities, yet they are a cost-effective solution for real-time data collection and observation in challenging terrain, remote areas, and among wildlife. One such sector that requires drones for safety and security is monitoring and surveillance. This includes the use of UAVs in day-to-day and emergency situations.

List of the monitoring and surveillance applications:-

1. Traffic Management
2. Crime Scene Investigation
3. Crowd Monitoring
4. Disaster Monitoring
5. Search and Rescue
6. Anti – poaching activities
7. Highway Patrol
8. Accident Site Investigation
9. Forest Monitoring
10. Criminal Tracking

Some of the above applications may seem interrelated; however, in practice, they differ from one another. Applications like - traffic management, crime scene investigation, and crowd monitoring- are related to police and security agencies. Whereas disaster monitoring, search and rescue, and anti-poaching activities are related to safety and lifesaving emergencies. They are interconnected but not necessarily interrelated, depending on the requirements.

6.1 Traffic Management

In populated countries, traffic is a big challenge, and traffic management is one of the toughest jobs. But this is not limited to populated countries only. Cities like London, Dublin, Toronto, Milan, Paris, and New York are among the most congested in the world, while cities like Bengaluru, Mumbai, Pune, and Beijing are among the most populous. Another problem in the traffic system is an accident, which can occur at any location, regardless of whether traffic is heavy or light.

UAVs can be part of smart cities, making them even smarter by implementing use cases in traffic management, safety, security, accident reporting, and urban planning. Moreover, national highways can also be considered for continuous patrolling.

6.1.1 Problems and Challenges

One of the major problems the world is facing is traffic congestion. Previously, traffic jams were an issue only in densely populated cities. However, with time, there was a significant increase in the number of vehicles and rapid population growth, i.e., users driving through the streets. Traffic mainly results from vehicles, but certain conditions can also cause long traffic jams. This includes old, narrow roads, improper roadside parking, and sudden lane changes, which cause traffic behind to halt. Accidents and work in progress are additional factors contributing to congestion.

The aforementioned reasons, along with all other factors leading to these conditions, also increase overall pollution, viz., air and noise pollution, as well as fuel consumption.

6.1.2 Drone as a Solution

UAVs capture data systematically and consistently wherever and whenever required. The first important thing done with a drone is real-time monitoring of the situation. This will provide an aerial perspective on the real-time traffic situation and the road-space coverage of all streets impeding traffic flow. Additionally, under this live monitoring, data on lane changers and their effects on traffic can be analyzed, along with the capture of live accidents. With live incident recordings, emergency services can be dispatched to the accident scene without delay.

Image 6.1 – UAV being Used in Traffic Management[73]

The traffic is always dynamic, and everyone's route is different because destinations vary. Under the held-up condition in a jam, UAVs can analyze the situation and direct traffic to alternative routes to avoid adding to the jam and worsening the situation. This can be done with drones equipped with onboard artificial intelligence that can analyze and identify routes and send the information to the control room while flying continuously.

6.1.3 UAV System Payload and Flight Operation

Various UAV systems are carrying out multiple operations. Among all, multirotor drones are the best fit for traffic management applications. The advantage of a multirotor drone is that it can hover in place to monitor the target. The preferred type of multirotor UAV is a quadrotor system, as it can be flown in narrow street areas. As per the requirements, the drone needs to be equipped with sensors and cameras.

A high-resolution RGB camera with a zoom facility is preferred for detailed visual monitoring. These cameras are gimbal-mounted and can be rotated to certain degrees in all three axes. A similar night-vision camera is recommended for nighttime monitoring. Certain UAVs have the capability to swap payloads, allowing them to be replaced with the required one as per the mission requirement. One step ahead, a combined sensor with day-and-night vision capture capability is recommended over two separate cameras. That will reduce the hustle of changing cameras twice a day.

The advanced version of the drone will have an AI-based system on board. This will help collect data by selecting a target in each frame for any illegal activity during the capture. Another advantage is that it can guide the control room with smart suggestions for alternative routes for commuters as part of traffic management. This works based on a model's machine learning process. In cases where heat signature detection is required in low-light conditions, thermal cameras can be added to the drone. However, this is required only in certain cases. For real-time data transmission, it is recommended that the drone have a long-range communication facility with 4G/5G network.

6.1.4 Data Acquisition and Processing

UAVs are vehicles that can fly manually and autonomously. The autonomous mission is a waypoint-based navigation method useful in situations where consistent traffic monitoring is required. (The details of preparing a waypoint navigation from one location to another is covered in the chapter - "Application-based mission planning for UAV operation"). A continuous waypoint mission is recommended when the requirement is to monitor a particular road or highway from a desired start point to an end point. This will follow a continuous mission.

The data collected from the drone can be viewed live by the traffic control room. This will help them make decisions on alternative routes, guide the ground staff on actions to take, and inform further planning to reduce traffic congestion, including changes to on-ground policy and infrastructure improvements. The collected data can also be used to train a machine learning model to analyze traffic patterns. Traffic predictions and alternative optimized routes can also be identified easily with AI.

6.2 Crime Scene Investigation

Crime scene investigation is a process of examining the crime site which includes the barricading the area, collecting photographs with close up, wide angle and macro shots, collecting evidences like fingerprints, footprints, weapon, blood, tire tracks, etc., tagging the evidences and getting the collected evidences analyzed by a forensic lab, recreating the scene, and establish connection between the victim, witness, and suspect.

Drones have become an invaluable tool in various applications, and crime scene investigations are no exception due to their ability to provide rapid, cost-effective, and detailed aerial views.

6.2.1 Problems and Challenges

Crime may occur at any location, including cliffs, dense forests, rooftops, resorts, large open areas, riverbanks, and very remote locations that are difficult for humans to reach. There are locations where manual investigation is difficult, but can be carried out. However, such locations can be tough and dangerous, and they may take a lot of time. Even then, there is a chance of missing tiny details because it is almost impossible to scan every centimeter of the area. Additionally, certain locations may be risky and hazardous due to forest animals, chemical spills, unstable structures, etc.

6.2.2 Drone as a Solution

UAVs can scan typical locations such as cliffs, dense forests, rooftops, resorts, large open areas, and the surrounding areas of a river, covering large areas in a few minutes. Drones collect high-resolution data using advanced cameras, which helps analyze every centimeter of the area and create 3D reconstructions. Drones can be equipped with thermal cameras that measure temperature, helping identify objects that are hidden or camouflaged to humans. With UAVs available in various sizes, it is possible to scan areas comfortably, even between the trees, by flying at low altitude.

6.2.3 UAV System Payload and Flight Operation

Crime scenes can best be captured using multirotor drones, especially quadcopters, which are small, stable, and can fly in tight spaces. These drones can be equipped with multiple cameras so investigators do not miss any evidence. A high-resolution RGB camera is required to capture detailed images and video. A surveillance camera with zoom capability is required to take a close-up look at the target. To identify heat-bearing signatures such as objects and footprints, a thermal imaging camera is required.

Such drone missions are carried out manually and autonomously. Manual flights are helpful for searching through the area using the drone's live data feed at the ground control station. The sensors used during manual flights are primarily surveillance sensors with zoom capabilities. This sensor requires the gimbal to be attached to the camera. Autonomous flight missions can be carried out using waypoint navigation and a complete grid mission, depending on the crime location and the requirements. Waypoint navigation helps investigators scan a particular area of interest. Whereas the grid mission will scan the complete area in a zigzag pattern. The area being scanned can be viewed live over the ground control station. Additionally, if an FPV-equipped drone is used, the team can wear goggles, fly manually, and view the live data directly in the goggles, giving the investigator a bird's-eye view.

6.2.4 Data Acquisition and Processing

Drones capture images and videos that are always stored on an onboard memory card, which can be viewed and analyzed after the data is transferred to a computer.

In the processing phase, the collected videos are manually analyzed in depth, and the resulting images can be used to create a 3D model of the site. The data collected by a thermal camera must be analyzed using temperature-measuring software.[74]

6.3 Crowd Monitoring

Managing a local fair, concert, social gathering, election campaign, etc., is typically easier than difficult when it comes to controlling the crowd. There are certain reasons for performing crowd monitoring. For events like election rallies, concerts, etc., there is a chance of fans running and pushing, and, in the end, a stampede. There are places that can accommodate only a few thousand people. With the counting algorithm, it is known that the limit has been reached, and the doors can be closed. This is mainly useful for managing the crowd.

6.3.1 Problems and Challenges

The traditional methods of observing crowds from a watchtower have limited visibility in dense crowds, making it difficult to identify bottlenecks. Manual watching relies on human-based observatory data, which delays decision-making at a crucial time to take action. Monitoring large crowds from the ground can place personnel in confusing situations, especially in volatile or rapidly changing scenarios.

6.3.2 Drone as a Solution

Drones are dynamic objects that can fly at any desired altitude, capture high-resolution images and videos from any required angle, focus on any particular target, and provide real-time data to the personnel and the command centers. This makes monitoring the crowd gathering safe and secure, cost-effective, and reduces the overuse of manpower.

By comparison, UAVs revolutionize crowd monitoring by overcoming its limitations. UAVs are not fixed observatory devices, so they can cover almost anything in the crowd with the help of high-definition sensors, changes in altitude and angle, and a focused area, providing real-time data.

6.3.3 UAV System Payload and Flight Operation

For large or small-area crowd gatherings, the most feasible solution is to fly a multirotor drone, preferably a quadcopter. Flying over the crowd is risky in case of any failure in drones; however, the lighter the drone, the safer it will be under such conditions. Considering the same, the lighter quadcopter with a high-resolution, zoom-capable camera system will be the most useful piece of equipment. Thermal cameras can additionally be used to detect any hidden weapons.

6.3.4 *Data Acquisition and Processing*

UAVs for such applications can be flown in both autonomous and manual modes. Flying within a targeted boundary area, setting different navigation points, and repeating it can be easy when combined with different flying altitudes. In between, the user can take over and fly the drone manually to the desired location.

The data collected can be observed in real-time on the command-and-control center screen. However, it will be saved on the drone. This data can then be manually monitored to identify whether any incidents or suspicious activities have occurred.

6.4 Disaster Management

Disasters are of two types: natural and man-made. The disasters that take place due to earthquakes, tsunamis, landslides, floods, forest fires, cyclones, etc., are natural disasters. Human-made disasters include industrial explosions, intentional fires, structural failures, etc. As disasters can take a serious toll on people and property, it is necessary to follow the disaster management process, which involves a systematic approach towards disaster prevention, preparedness, response, and recovery. Disasters are emergencies where a slow response can be dangerous. But humans do have their limitations when responding to disasters. The safety of disaster relief forces is the first priority; only then can they safeguard the humans affected by the disaster. UAVs are very helpful in all four processes of systematic disaster management, starting from prevention to recovery. Disaster management is mostly required for the following disasters:-

a. Heavy rain and flood situation

b. Landslide

c. Forest fire

d. Cyclones/hurricanes

e. Tsunami
f. Earthquakes
g. Droughts
h. Infrastructure failure
i. Hazardous material exposure
j. Nuclear explosion
k. Epidemic
l. and much more

6.4.1 *Problems and Challenges*

The existing monitoring and surveillance techniques include satellite and helicopter surveillance of the disaster site. Helicopters have limited access to the areas due to difficult terrain, damaged infrastructure, persistent bad weather, etc. Satellite imagery may lack the required resolution due to cloud cover, preventing an understanding of the ground situation. Though nothing is as useful as satellites in terms of getting information about the changes in climatic conditions in advance. Sending the ground teams to the disaster zone to gather data and share it with headquarters is risky and may delay the entire rescue and relief process.

6.4.2 *Drone as a Solution*

Drones are a cost-effective and quickly deployable solution for post-disaster activities. Flying to disaster zones and providing real-time data can be easily done using drones. Flying a few meters above the ground in terrain-following mode, it can capture high-resolution data and send it to the ground team. At locations where communication towers have been destroyed, drones can fly and provide a temporary network for users in affected areas. There are different ways in which drones can be utilized in disaster management.

A. **Prevention** – Disaster prevention is the key requirement, and this is where drones are useful. Not all disasters, but some can be prevented. For example, a lake wall can break if water overflows and the wall is weak. There are houses nearby that could be affected if such an incident occurs. A drone can collect data and understand the water flow analysis.
B. **Preparedness** – Drones can help teams understand the possibility of potential risks and relocate people to a safe place for a temporary period of time. This saves many human lives.
C. **Response** – During the calamities, drones can take off in a minute after the rain stops. Saving time and reaching the affected site in no time. It helps take action faster. Delivery drones are helpful for dropping food packets and first-aid supplies.
D. **Recovery** – Once the calamities are over, drones can help identify the vanished roads by comparing with the data collected before the incident. This way, the older route can be repaired and reopened for public use.

6.4.3 UAV System Payload and Flight Operation

There are different types of UAVs that can be useful in disaster management. Multirotor UAVs are best suited for surveillance and data collection. The suggested payload for multirotor drones includes high-resolution cameras for photography and surveillance cameras with zoom capabilities for manual videography. Thermal cameras can help identify humans and animals with temperature differences. This can help in saving a human life.

For disasters where a long flight time is needed, and a straight stretch is available for takeoff and landing, fixed-wing drones can be used. It can fly for kilometers and collect enough data to be useful for multiple locations or large-area analysis. A surveillance camera with zoom capability and a high-resolution camera are suggested as payloads for fixed-wing UAVs.

Image 6.2 – Data Capturing of a Flood-Affected Area[75]

Hybrid VTOL drones can take off vertically and fly like fixed-wing aircraft. Hence, these drones can carry a much larger load, which can be delivered by landing at a site and taking off back to the base location. These types of UAVs are capable of carrying a surveillance camera along with a required payload to be dropped.

For all drones, manual flight is possible; however, for fixed-wing and VTOL drones, it is recommended to use autonomous flight with grid or waypoint navigation. Whereas for a multirotor, both manual and autonomous operations can be carried out depending on the requirements.

6.4.4 Data Acquisition and Processing

The collected images can be useful to create a 3D model of the site. This data will help analyze overall destruction by comparing it with data collected before the calamities. The collected videos are manually analyzed in depth. The data collected by a thermal camera must be analyzed using temperature-measuring software. A detailed report can be generated to help the ground team understand and easily decide on their next response to relief efforts.

6.5 Anti – Poaching Activities

Poaching of innocent animals is common in the majority of the forests across the world. The motive behind poaching is the collection of animal parts and selling them illegally. The majority of poaching occurs for the following purposes: the collection of tigers' and lions' nails, elephant ivory tusks, rhinoceroses' horns, rabbits, and other animals' fur, etc. Poaching is a threat to biodiversity by endangering species like tigers, rhinos, elephants, etc.

6.5.1 Problems and Challenges

The existing problems for any forest department are that poachers enter during the night from remote areas or dense forests. This makes it difficult to detect poachers, compounded by a lack of manpower and other technological resources. However, the existing solutions include ranger patrols and the setting of traps for poachers. Additionally, fixed cameras are also installed at critical locations. Radio collars are used to track endangered animals. Overall, all the above-mentioned solutions have certain limitations.

6.5.2 Drone as a Solution

Drones can be used for the detection and prevention of poaching activities, as well as for catching poachers. UAVs equipped with cameras and sensors, including thermal cameras, can patrol the areas efficiently while providing real-time data to the forest base offices. The common, easy entry points for poachers can be identified and actively monitored. Rangers can have real-time visuals from the drone, enabling them to reach the site or catch poachers. Between the dangers posed by humans (poachers) and animals, drones are an effective solution that reduces the risk for forest guards.

Image 6.3 – A Drone with an AI-Based Detection System[76]

6.5.3 UAV System Payload and Flight Operation

Fixed-wing UAVs can map the entire forest, and multirotor drones are the best option for day-and-night surveillance as well as thermal photography. A surveillance zoom camera with infrared night vision is useful for day and night monitoring. These drones can have a swappable payload system, allowing cameras and sensors to be interchanged as required.

6.5.4 Data Acquisition and Processing

Autonomous missions are designed for forest coverage, and they will map the entire forest. Waypoint navigation may be useful for repeatedly monitoring a particular location. Manual flying of the drones will be preferred for real-time tracking of activities. The majority of the data consists of video footage and infrared images. Video footage is to be analyzed thoroughly. Infrared images require software processing to detect temperatures and determine whether any humans were detected. The recorded data should be kept for record-keeping and reference, as it can be useful for identifying hotspots where poachers have been detected or where there are potential illegal entry points into the forest.

Points to Remember:-

- The majority of monitoring and surveillance activities can be carried out using only a 3-axis RGB zoom camera with night vision.
- Monitoring an industry can be done using UAVs in autonomous mode, which allows the drone to continuously surveil at night.
- The majority of the surveillance data, i.e., captured videos, does not require any data processing.

CHAPTER

07

UAVs for Outdoor Inspections

Unmanned Aerial Vehicles have the majority of use cases that require it to be flown in an open sky area. This includes applications ranging from mapping, surveying, surveillance, monitoring, inspections, spraying, and deliveries. These open-area applications always need a GPS system for precise location identification and returning to home locations. There are other outdoor applications beyond the few mentioned above.

The list of a few inspection applications requiring drones to fly in the open environment is:-

1. Solar power plant inspection
2. Wind turbine inspection
3. Oil and Gas pipeline leakage detection
4. Power line inspection
5. Transmission tower/telecommunication tower - asset inspection
6. Railway track inspection
7. Chimneys
8. Infrastructure Inspection
9. Refinery Inspection

These outdoor inspection applications are unrelated; therefore, different methodologies are used to execute the project. This process may require a variety of sensors, a separate mission-planning process, and multiple data-processing software tools.

Inspection of the above-mentioned and other applications is required to be conducted in accordance with the organization's SOP and local government regulations governing industrial safety and security. The inspection of railway tracks, oil and gas pipelines, and chimneys is related to the safety of the industrial workers and commuters. Inspections of powerlines and transmission towers are required to identify structural issues in the towers and powerlines and to calculate losses.

Renewable energy sources have gained significant importance in this decade, mainly due to the scarcity of coal, the main source of electricity generation in many countries even today. Other methods of electricity generation include natural gas, hydropower, nuclear power, wind turbines, and solar photovoltaics. Renewable sources, i.e., wind turbines and solar panels, depend on wind and sunlight. So, it is understandable that its energy generation capacity will vary from day to day. For the same reason, inspecting these renewable sources is important for maintaining high power generation efficiency.

The above-mentioned outdoor inspection applications are explained below. Each of these applications follows the same process from project initiation to the end result. This process is as follows:-

- Check weather – atmosphere and irradiance level
- Planning the mission
- Data acquisition
- Data processing
- Report generation

7.1 Solar Power Plant Inspection

Solar power generation has gained momentum and become a major source of renewable energy. The main benefit compared to a wind turbine is that its structure is lighter, making it easier to transport and install the fixed equipment. Additionally, it is easier to maintain than around 100-meter-tall wind turbine power generation units, and it is a cheaper way to generate efficient energy for supply. The rooftop panels in urban areas are rated in kilowatts, whereas solar power plants are rated in megawatts and gigawatts.

Image 7.1 – GW Solar Power Plant[77]

Renewable energy from solar panels is generated by the energy transmitted by the sun. These panels, formed of solar cells, absorb the energy and supply it to the respective supply board. However, it is always important to have a scheduled inspection of the entire farm to check connections between panels, broken glass on panels, faulty solar cells, dirt and dust on panels, periodic cleanliness, etc. These defects are the major drawbacks that reduce the overall efficiency of power generation.

For this purpose, an assigned inspection team will thoroughly go through the plant. This sort of manual inspection takes around a month or even more in large plants. Compared to this manual work, drones finish the same task in a few days, depending on the area. Overall, a 25 MW area can be covered in a day. The above-mentioned inspection for defect identification can be carried out using this method, and even more can be covered in a single inspection flight over the entire area. Once the data is captured, it goes to the data processing team to apply algorithms and detect defects. Finally, the report is generated and submitted to the clients.

7.1.1 Requirement of Solar Power Plant Inspection

A solar power plant is a combination of multiple parts and structures. A solar panel consists of silicon cells. These panels are laminated. Multiple panels are combined and connected as required to fulfill the intended purpose. There are connections, joints, and multiple points where defects may occur. These defects reduce the efficiency of energy generation.

Below is the list of defects that may occur in any solar farm:-

- Microcracks
- Delamination
- Manufacturing defects
- Overheating of panels
- Snail trails
- Busbar corrosion
- Improper wiring
- Poor tracking system installation
- Substandard tracking
- Inadequate foundation

The last four defects listed above are related to construction and can be eliminated in a single process. However, for solar panel defects, it is necessary to inspect them periodically.

7.1.2 *Payload Required*

Solar farms are where the heat energy plays a major role. This is why the sensor required here is a temperature detecting sensor cum infrared camera. This type of payload captures infrared images for identifying defects, such as hotspots, panel performance assessment, and electrical failure detection. These are the key parameters that must be checked thoroughly at regular intervals to maintain efficiency and improve power generation performance.

To identify the precise solar panel locations and generate an inspection report, regular photo and video documentation is required. This can be done with an RGB visible light camera.

There are many thermal sensor manufacturers producing only thermal cameras, thermal + RGB cameras, and 3-in-1 combo camera sensors, as well as gimbal stabilization. This is a combination of thermal and RGB cameras in a single unit, with 2- or 3-axis gimbal control. This reduces the requirement of two separate drones for two different sensors. Additionally, it reduces the total number of flights required to cover the entire area and the hassle of switching or attaching two different payloads. A separate camera will require a gimbal as an add-on facility for attachment.

Thermal infrared cameras are highly sensitive and can detect even subtle temperature variations. Its other capabilities include a wide temperature detection range and detailed image resolution for accurate analysis.

Image 7.2 – Siyi ZT6 Mini[78]

The above image shows an RGB visible-light camera and a thermal imager combined in a 3-axis, stabilization-supported gimbal.

7.1.3 Data Acquisition and Processing

Planning a mission saves a lot of time in the field when operating within a limited timeframe. Certain information is required before planning any mission, such as the site location, the area or farm to be covered, etc. For mission planning for solar panels, one of the ground control station software options is Litchi, among others. However, the Fly Litchi application is only for DJI brand drones and does not support other drones with open-source software. Other GCS platforms used for autonomous flight include the Q Ground Control Station (QGC) and the mission planner. Other application-based platforms can also be used for the same purpose.

Planning an area

Select the area that needs to be inspected. Once the area to be covered is selected, grid plans are made over the solar panel with an overlap. This happens in such a way that the area before the solar panel starts and after it ends will also be covered. Overall, each image will have a 50% overlap on the sides and the front. Check the image for the reference.

Image 7.3 – Grid Mission Planned Over a Solar Panel

Considering the use of mission planner software for this example. While planning, a polygon must be created. This is the area of interest over which the drone flight/mission is to be carried out. Accordingly, the mission will be conducted in a grid formation, with the drone flying from one end to the other, capturing images along the way. Grid missions are a zigzag path (as discussed in the mission planning chapter). Drones are programmed to follow this path autonomously. Once the target area is selected, a grid must be generated. Once this is done, a mission identical to the one shown in the image above will be generated. The layout shown above, or the generated path on the screen, is not final.

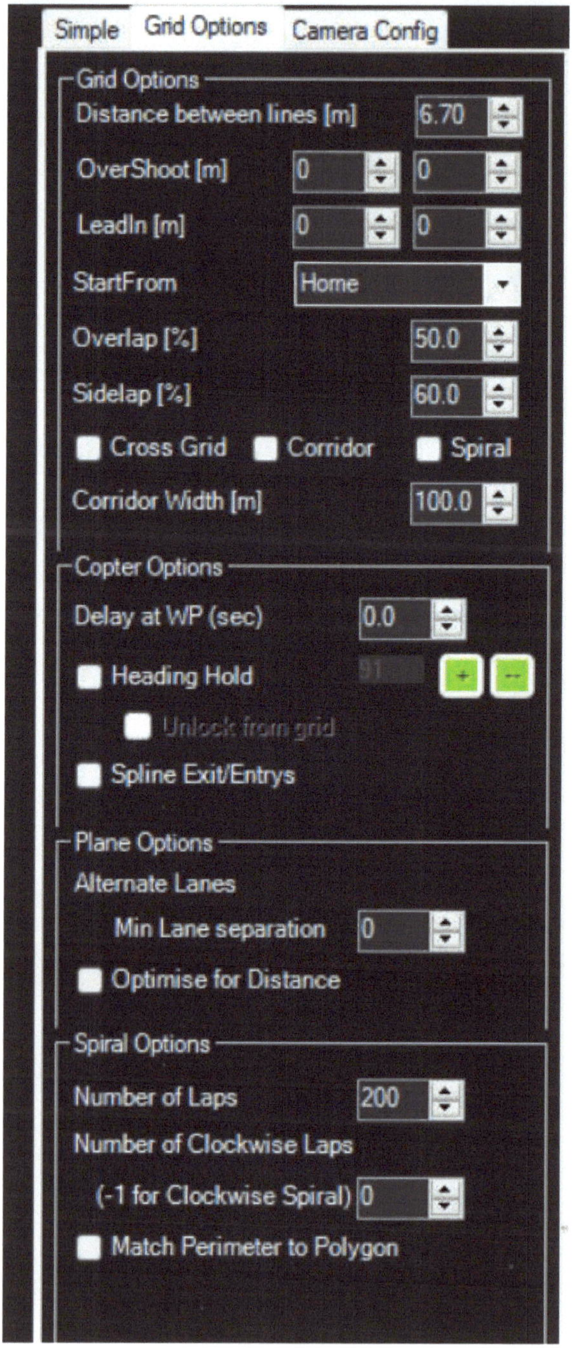

Image 7.4 – Mission Planning Process on Mission Planner Software

This is a basic planned mission that requires the user to adjust parameters such as camera selection, flying altitude, grid angle, image overlap settings, line spacing, and camera angles. These are common settings, among others, that need to be set per the mission requirements. The image above shows a few important settings that may need to be set while planning and adjusting a mission.

The flying altitude for solar panel inspection depends on various parameters. One, the size of the panel. Two: the camera resolution and the desired GSD. However, the altitude ranges from 15 meters to 40 meters.

The captured thermal images need to be processed using specific applications, such as FLIR systems. This application is useful for analyzing all images to identify hotspots, faults, and other features. Other thermal processing software is also available.

Key Deliverables and their objectives:-

Output	**Purpose**
RGB – Ortho mosaic	– Panel layout verification – Structural damage identification
Thermal – Ortho mosaic	– Hotspot detection – Faulty panel identification
Defect Reports	– Maintenance prioritization

Table 7.1 – Solar Inspection Outputs

7.2 Wind Turbine Inspection

Wind turbines are tall structures with blades that convert kinetic energy into electrical energy. The typical height of such a wind turbine ranges from 60 meters to 120 meters. Considering major key factors:- safety, time required, reliability, and cost effectiveness, drones provide a comprehensive and efficient solution for wind turbine inspection. Drones can collect detailed data, enabling proactive maintenance of wind turbines and optimized energy production.

Image 7.5 – Image of a Wind Turbine Using a Drone[79]

7.2.1 Requirement of Wind Turbine Inspection

Wind turbines are tall structures installed mostly in open areas where the wind blows continuously. Because of this, the structure faces weather-related effects, including dust storms, rain, cyclones, and other climatic changes, which can damage it and reduce power generation efficiency.

Below is the list of defects that may occur in any wind turbine:-

- Leading edge erosion – reduces aerodynamic efficiency
- Surface cracks – material defects
- Impact damage – due to hail storm, bird strikes, dust storm, etc.
- Separation of blade layers
- Corrosion
- Fatigue cracks
- Component overheating

7.2.2 Payload Required

The majority of the defects can be identified using an RGB camera for photography and videography, or a zoom camera. Often, LIDAR sensors are used to measure the precise geometry of the blades and the dimensions of the wind turbine. It is easier to detect structural deformation and imbalances with LIDAR than with other sensors.

There are still a few surface defects that are better identified with thermal infrared cameras. This includes capturing defects such as blade defects and component overheating.

Here is a list of payloads that will be required to fulfill the above requirements for the identification of the defects:-

a. 10x to 30x zoom camera

b. RGB camera

c. LIDAR sensor

d. Thermal camera

7.2.3 Data Acquisition and Processing

The round vertical structures are typical in mission planning and data capture because they need to be inspected vertically, in a straight line, from every angle. Additionally, all the parts must be inspected thoroughly in a layer-by-layer form.

For planning the autonomous inspection of such structures, there are different applications. One such application is "Hammer Missions". This application covers almost all the parts and portions of the wind turbine.

However, the Hammer Mission application only supports DJI-based drones. Support for PX4, ArduPilot, etc., will soon be launched.

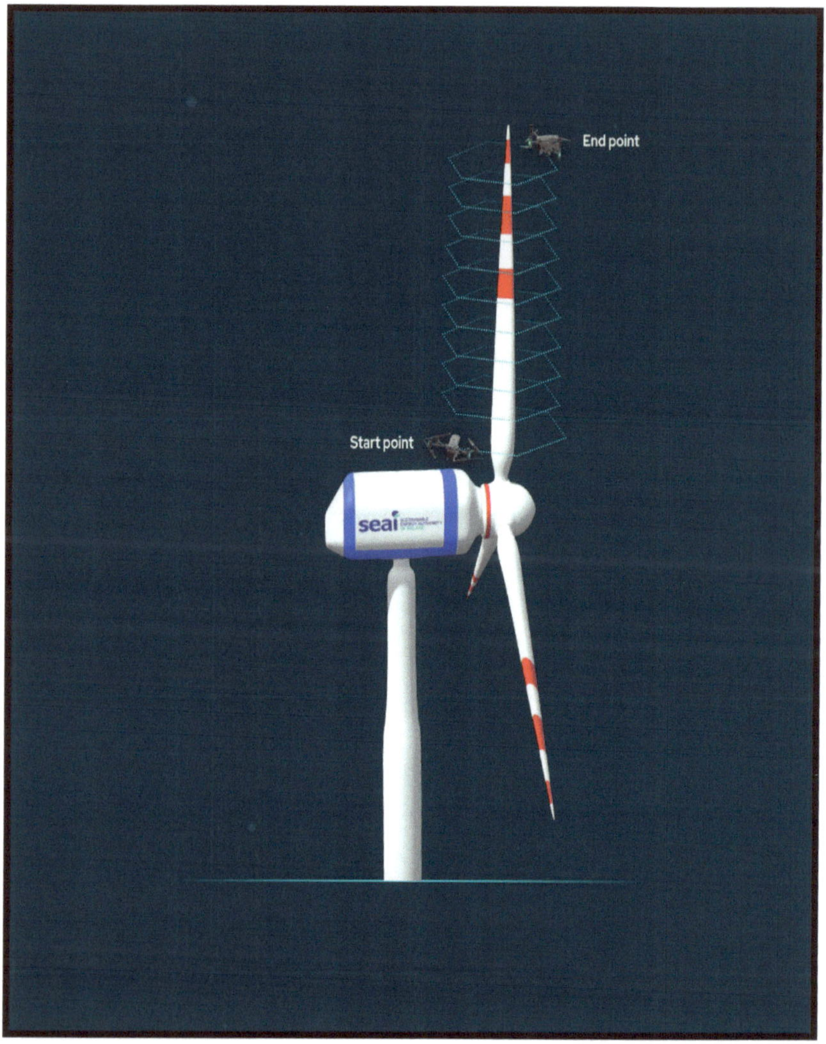

Image 7.6 – Inspection Using the Hammer Mission Application [80]

Another GCS that allows a structure scan is the Q Ground Control Station application. This structure is easy to plan and executes the mission in a layer-by-layer form. This scanning is based on various parameters for its autonomous mission. The key parameters for this include scan distance, layer height, trigger distance, structure height, and scan bottom altitude.

Image 7.7 – Planned Mission for Structure Scan on QGC

The above image shows the planned autonomous inspection of a wind turbine. The altitude is set at 100 meters. The scanning distance is considered 10 meters. Scanning distance is the distance from the drone to where the image will be captured. Layer height is the altitude gap between two horizontal scans by the UAV. Layer height is relative to scan distance and will vary accordingly. The height of the structure considered here for a wind turbine is 100 meters.

The altitude at which the lowest layer of scanning occurs is the Scan Bottom Altitude, set at 10 meters above the ground. Triggering distance is the interval distance between two consecutive images captured. This autonomous mission will last 16 minutes and capture approximately 1,000 images. The entrance and exit altitudes can also be set based on the structure and mission convenience.

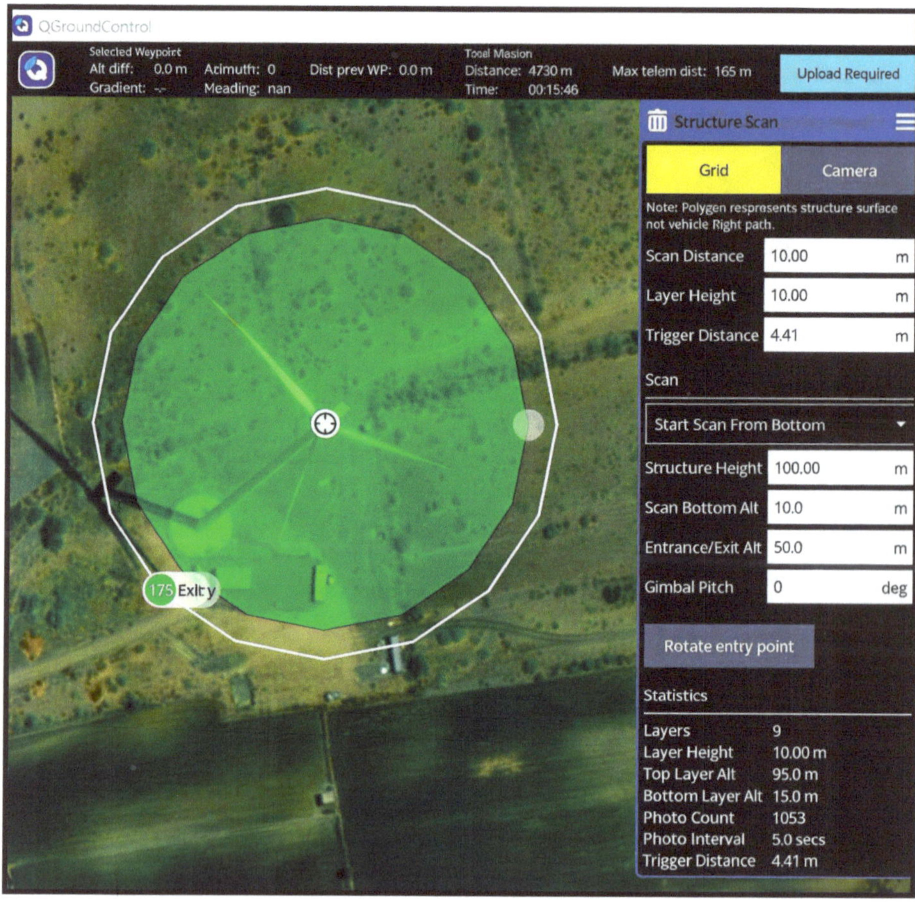

Image 7.8 – Details of the Set Parameters for Structure Scanning

Key Deliverables and their objectives:-

Output	Purpose
High – Resolution RGB Images or Zoom camera	– Crack, erosion, and surface defect identification (Mostly included in final reports)
Thermal Images	– Overheating and internal fault identification
Inspection Reports	– Maintenance and safety planning

Table 7.2 – Wind Turbine Inspection Outputs

7.3 Powerline Inspection

In densely populated countries like India, power transmission and distribution lines total approximately 14 million circuit kilometers (at the time of writing this book). The next fact is that it is difficult to inspect millions of kilometers of power lines manually, given varying climatic conditions, terrain, forested areas, rivers, heavily trafficked roads, etc. The manual inspection process will be too slow to increase the overall efficiency of electricity transmission. It is even difficult to estimate the timeline for manually inspecting the entire 14 million km of powerline. The time spent in transportation, critical methods of climbing, and then inspecting, which is then reported. This requires a lot of effort, time, and high expenses. On the other hand, drones are capable of executing this process quickly and accurately. It is safe as long as humans remain on the ground. Additionally, UAVs can reach places that are difficult to access for manual inspection.

Image 7.9 – Powerline Inspection from a Drone[81]

7.3.1 Requirement of Powerline Inspection

Powerline failure has a significant impact on society as a whole. Apart from other defects, there are certain risks associated with it, including fire due to a short circuit. The list of defects that may occur through the power line is listed below:-

- Corrosion in conductors
- Wear and tear of conductors
- Loose strands
- Partially broken conductors
- Flashover
- Contamination of the insulator
- Cracks and breakage of the insulator
- Vegetation interference
- Lightning strikes
- Other impacts

"A power cut is better than a power failure."

7.3.2 Payload Required

Under the powerline inspection project, either multiple drones with different sensors and cameras on board will be needed, or a single drone capable of carrying all your payload requirements will be required. This can be turn-by-turn or an attachment on a single drone.

Vegetation encroachment over the pole and foundation can be detected using a multispectral sensor. Using this multispectral band photography, insulator contamination can also be detected. A high-resolution visible RGB camera can capture detailed images of the powerline infrastructure mapping. With an RGB camera, visible defects can be identified, such as wear and tear, corrosion, and structural damage.

Image 7.10 – Thermal Imagery of a Transmission Line[82]

Using a thermal camera, hotspots and fire spots, conductor heating, etc., can be detected. This allows users to identify even a minute temperature variation in any portion of the line. Apart from this, LIDAR sensors are helpful in creating 3D measurements of the structure. This allows it to accurately inspect defects such as sag and clearance issues. When equipped with a high-quality zoom camera payload for visual inspection, there may be no need to switch off the power supply via the lines.

7.3.3 Data Acquisition and Processing

For powerline inspection and mapping, one of the best ways to plan your capture is to create a straight, linear path from one end to the other. The area covered by powerlines is in a long stretch with almost no width. So, the area required to be covered in width is less. So linear path planning is the best approach.

In the linear path approach, the drone will fly along a straight line generated by the user. The flying altitude will vary as per the user's requirements. However, the entire target area can be covered by a single linear, planned path along the line. Another option for planning a mission is waypoint navigation. In this method, multiple points can be added to the plan and can be executed by the user. At each point, the altitude can be defined, and all points will be connected, forming a single line. The above-mentioned methods are for static sensors such as RGB camera, thermal camera, and Lidar, where zooming is not available. When using the Zoom camera system, flying can be done manually by holding a particular altitude. This way, the pilot can control the drone, stop it at the required position, and zoom in on the target area for visual inspection.

Both waypoint navigation and a linear path/corridor can be executed autonomously without any human intervention. Under waypoint navigation, the end point must be defined without failure.

(For corridor mission planning methodology, please refer to the railway track inspection in this chapter)

Key Deliverables and their objectives:-

Output	Purpose
RGB Images	– Physical damage identification
Thermal Images	– Hotspot detection – Overload detection
LIDAR Point Cloud	– Line sag calculation – Vegetation clearance analysis
Zoom Camera	– Visually inspect and mark the defects

Table 7.3 – Powerline Inspection Outputs

7.4 Transmission Tower Asset Inspection

Transmission towers are critical infrastructure, always located in open areas, loaded with high-voltage extension cables used to transmit power. These structures, when placed in open areas, are affected by the high voltage of the extension line, climatic conditions, and continuous stress from other environmental factors. As a result, it is necessary to inspect the same at regular intervals.

A manual inspection of the transmission tower will require a scheduled power cut. The team climbs the tower, checks for any irregularities, replaces the part (if any), and gets down. With drones, the requirement for scheduled power cuts can be reduced by at least by 50%. The first reason is that drones can inspect assets without requiring a power cut. When the inspection is complete, the list of all towers with various defects, along with information on specific parts replacement/repair, is required. This information is collected without shutting down the line, saving the field team time and effort and resulting in higher work efficiency and lower expenditure.

Image 7.11 – Transmission Tower Asset Inspection[83]

7.4.1 Requirement of Transmission Tower Asset Inspection

Transmission tower needs to be strong and free from defects, because any critical damage to a tower may lead to a major power failure. To prevent undetermined major power failures, it is better to inspect the towers at regular intervals and address any deterioration or defects.

The following defects are commonly identified on a transmission tower:-

- Material corrosion
- Cracks
- Welding joint defects
- Nut-bolt failures
- Erosion and tower foundation defect
- Lightning strike damage
- Fire

7.4.2 Payload Required

Transmission towers are lattice structures standing at around 40 meters above the ground. Flying drones near and around the transmission tower is critical. A 20 MP RGB camera can be used to capture the detailed visuals, including cracks, corrosion, and bolts. Along with this, a Zoom camera that can capture clear video from a short distance for safety can also be considered for this purpose.

For deformation analysis of the tower structure, a 3D model created from a LIDAR survey is useful. Thermal cameras can detect temperature variations, hotspots, and overheating of the parts.

7.4.3 Data Acquisition and Processing

For image capture or videography using a zoom camera, the inspection is carried out manually while maintaining altitude by flying along the sides of the tower and inspecting all parts as needed.

For drones with thermal and RGB cameras, autonomous mission planning can be carried out using parameter settings for different flying altitudes, and the required camera angles are set accordingly. For an autonomous mission, it is beneficial to have an RTK system on board the drone to achieve higher horizontal and vertical positional accuracy.

Key Deliverables and their objectives:-

Output	Purpose
High Resolution RGB images	– Component – bolt, insulator, etc., inspection
LIDAR Point Clouds	– Structural geometry and clearance analysis
3D Models	– Digital asset documentation
Zoom Camera	– Visually inspecting the defects

Table 7.5 – Transmission Tower Inspection Outputs

7.5 Railway Track Inspection

Railways are one of the most important transportation systems in the World. In Railways, delays are crucial yet ignorable compared to incidents or accidents caused by unidentified defects. Even today, in certain countries like India, railway tracks are manually inspected on a day-to-day basis. A person travels along one track from one end to the other, looking at both sides to identify any damage, defects, or other issues. And the same happens with other track inspection procedures. This entire process is safe but time-consuming and requires a lot of effort to walk through the tracks. Compared to that, UAVs require much less time to inspect the required track and generate its results, precisely while flying above the track.

Image 7.12 – Railway Track Inspection[84]

7.5.1 Requirement of Railway Track Inspection

To ensure safe transportation, it is essential to inspect the railway tracks. Below are the defects that occur on railway tracks:-

- Head checks
- Split head – vertical crack extending through the rail heads
- Transverse fissure
- Horizontal fissure
- Sudden fractures
- Corrugation
- Joint defects
- Wear and tear
- Deformation in rail heads
- Ballast fouling
- Track geometry defects
- Sleeper defects
- Fastener failure

Image 7.13 – Common Rolling Contact Fatigue (RCF) Defects (Head Checks, Shells, Flaking, Thermal Cracks)[85]

7.5.2 Payload Required

Critical surface defects can be detected using high-resolution CMOS sensor-based RGB visible-light cameras. These defects include split heads, head checks, transverse fissures, corrugations, etc. To measure track geometry, identify soil erosion, etc. LIDAR sensors are useful.

7.5.3 Data Acquisition and Processing

For Railway track inspection, one of the best ways to plan your capture is to create a straight, linear path from one end to the other. The width of the area of interest, the railway tracks, is typically a few meters. The area increases when multiple tracks are present, most often in the vicinity of a train station.

In the linear path approach, the drone will fly along a straight line generated by the user. The flying altitude will vary as per the user's requirements. However, the entire area can be covered by a single linear, planned path along the line. Another option for planning a mission is waypoint navigation. In this method, multiple points can be added to the plan and can be executed by the user. At each point, the altitude can be defined, and all points will be connected, forming a single line.

Image 7.14 – Corridor Mission Planned for Railway Track Inspection

The image above shows a planned corridor mission over a pair of railway tracks. For better understanding, a 1.5 km stretch is considered, with a width of 20 meters. The drone's flying altitude is set to 50 meters above ground level. This much area can easily be covered with a drone flying in the center of the corridor width without repeating any path.

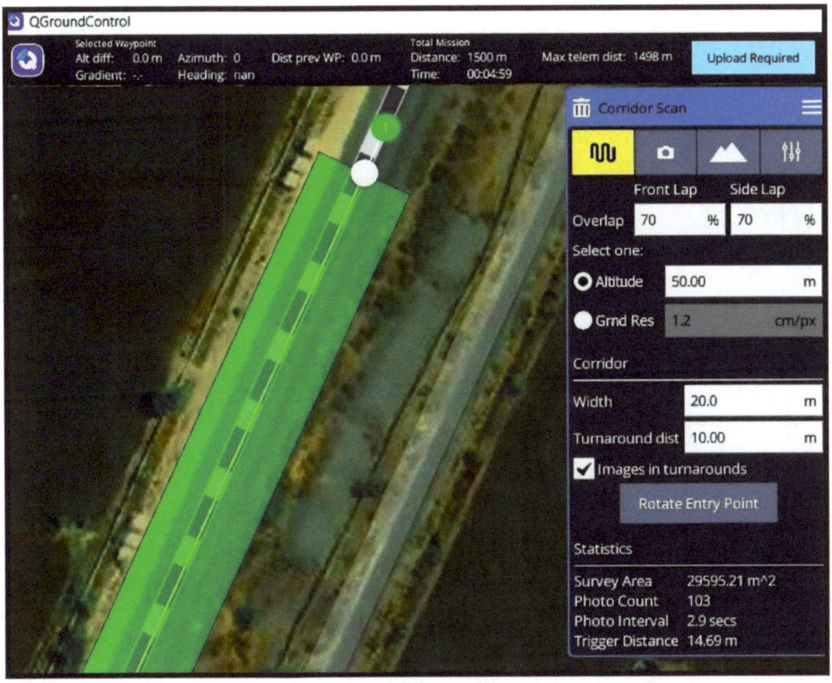

Image 7.15 – The Details of the Planned Mission Over a Railway Track

The above image shows zoomed-in details of the flight plan. The planned green area is the area of interest. The white line in the center, with point ❶, is the path the drone will fly. The UAV will take approx. 5 minutes to cover the complete area and capture around 100 images with 70% front and side overlap.

Key Deliverables and their objectives:-

Output	**Purpose**
Ortho mosaic	– Track geometry – Encroachment monitoring – Fault detection
Lidar Point Cloud	– Track level analysis – Cant and clearance analysis

Table 7.6 – Railway Track Inspection Outputs

Apart from traditional methods of defect identification, there are also systems available for analyzing drone-captured data, which are fully automated and do not require manual detection of defects. These AI-based SaaS systems are capable of detecting rail gauge, broken rails, gaps in track joints, cross-tie separation, ballast coverage, fastener presence, etc.[86]

7.6 Chimney Inspection

Chimneys are structures designed to safely remove smoke, gases, pollutants, and other emissions from factories and plants. The identification of corrosion and cracks on the chimney is typical, unsafe, and time-consuming when done manually. There is a risk to the worker's life who climbs the chimney, collects the note, and returns. Additionally, this manually collected data is not accurate. Drones make the chimney inspection process faster, more accurate, and safer. This leads the industry to carry out inspections as per the schedule without any risk. Chimneys must be inspected for maintenance from both the outside and the inside.

Image 7.16 – Chimney Inspection Using a Drone[87]

7.6.1 Requirement of Chimney Inspection

Defects in industrial chimneys can pose significant risks to both the environment and human safety. These defects can arise from a variety of factors, including construction errors, material degradation, operational stresses, and environmental factors:-

- Cracks
- Breakdown of lining material
- Corrosion
- Obstructions in the flue
- Spalling – surface deterioration
- Erosion due to wind, rain, and chemical exposure
- Deformation of the chimney
- Damage to ladders

7.6.2 Payload Required

Identification of visible defects, such as cracks, spalling, and erosion, can be captured using a high-resolution RGB camera. LIDAR sensors are good at precise measurements, so they can be used to create 3D models of the chimney while detecting additional defects such as cracks, settlement, and buckling.

7.6.3 Data Acquisition and Processing

To inspect the round vertical structures, a system specifically designed for circular object inspection must be used. These are typical structures for mission planning and data capture because they need to be inspected vertically in a straight line from every angle. Additionally, all the other parts must be inspected thoroughly in a layer-by-layer form. For details on setting parameters for autonomous structure scanning, refer to the data-capturing methodology section on wind turbine inspection in this chapter. The hammer mission solution can be applied to a chimney, as to any other vertical structure.

Also, the QGCS application can easily scan vertical structures using the structure scan method. This scanning is based on various parameters for its autonomous mission. For this, the major parameter settings include scan distance, layer height, trigger distance, structure height, and scan bottom altitude, all of which must be set.

Key Deliverables and their objectives:-

Output	Purpose
Lidar Point Cloud	– Verticality and deformation analysis
RGB Images	– Surface crack and corrosion detection

Table 7.7 – Chimney Inspection Outputs

Points to Remember:-

- For utility inspections from the outside, the selection of GCS and mission flight parameters must be carefully set. A small mistake may damage both the costly equipment and the drone.
- Always try to optimize the flight in the best possible way when using multiple sensors across multiple drones, so that the day's overall efficiency remains high.
- While inspecting power lines and transmission towers, fly drones by keeping a safe distance from the live cables.

CHAPTER

08

UAVs for Indoor Inspections

Inspections are, in fact, an observation of the targeted structures whose quality, material, strength, and other characteristics may deteriorate over time. Industries operate on continuous working hours, whereas manual inspection requires significant time and effort, making it difficult for the industry to balance the selection and timeline of periodic maintenance with continuous operation while maintaining efficiency. Additionally, certain machines and tanks are impossible for a human to reach for inspection. This further makes it difficult for the industry to maintain its entire system to the required quality standards.

List of Indoor inspection applications:-

1. Boiler inspection
2. Chimney and stack inspection
3. Structural integrity checks
4. Storage tank inspection
5. Sewer and drain inspection
6. Tunnel inspection
7. Underground mining inspection
8. Under the bridge inspection
9. Infrastructure inspection

UAVs designed and developed for indoor industrial inspection applications are fully proven systems that can enter almost any industrial space, such as boilers, to capture data. These drones are well-equipped with cameras/sensors and LED lights. Drones can fly to any height within the industry, reducing or even eliminating the need for scaffolding. Hence, it reduces manual effort and time required to stop the operation during maintenance hours. The system requiring maintenance now works better and, in turn, increases the efficiency in all three ways: time and work.

These drones, known as confined-space drones, are specifically designed UAVs with a protective cage and high-intensity lighting for inspecting enclosed, GPS-denied, and hazardous environments that are impossible for humans to enter due to the difficulty or risk. These drones with protective cages are not only keeping humans safe but also keeping themselves and the plants safe.

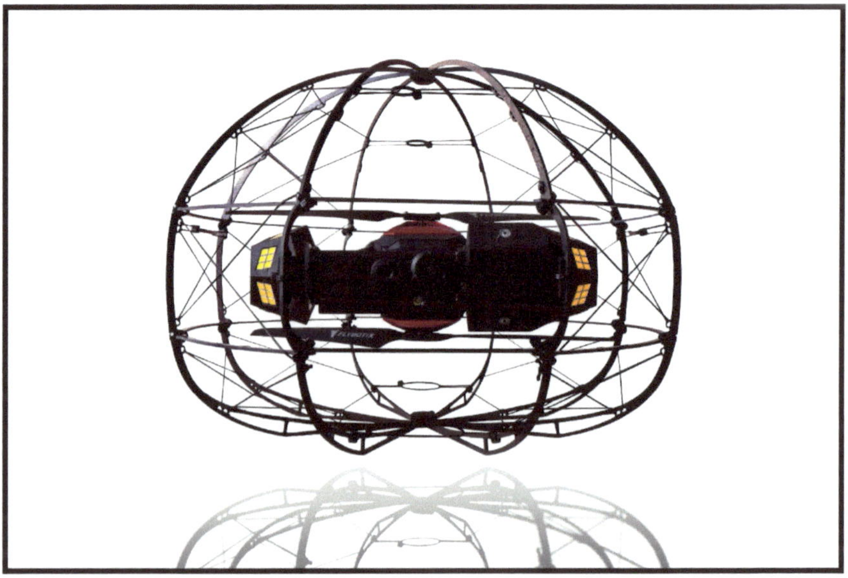

Image 8.1 – Flybotix ASIO-X Confined Space Drone[88]

The above is a Swiss-made ASIO-X confined-space drone model from Flybotix. This robust cage drone flies stably for approximately 20 minutes and is equipped with 40K luminescence light, a gas detector, and an LIDAR sensor.

8.1 Boiler Inspection

Boilers are water-containing closed vessels used to generate steam for power generation, chemical processing, wastewater treatment, etc. Varieties of industry like oil refineries use it for heating crude oil and for other chemical reactions; the food processing industry pasteurizes and sterilizes using boilers; the textile industry requires it for dyeing, drying, and finishing fabric, whereas paper mills use it for drying and processing paper; and for turbine driving in power plants.

Boilers are important in the above industry mainly for their consistency, as they provide a reliable source of thermal energy for long-duration operation. It helps maintain safety and quality standards in the food processing and pharmaceutical industries. The versatility of its operation across various fuels, such as oil, coal, and even biomass, makes it a better choice. There are different types of boilers made as per industry requirements: fire-tube boilers, water-tube boilers, electric boilers, biomass boilers, etc.

Image 8.2 – Industrial Steam Boiler[89]

8.1.1 Requirement of Boiler Inspection

Inspection of a boiler is an examination of its body, both internally and externally, covering components such as tubes, walls, burners, refractory linings, and welding joints. This periodical inspection is carried out to check the wear, corrosion, deformation, and cracks in the boiler. These inspections are very important as they are always in operation at high temperatures and pressures. Inspecting at a regular interval not only extends the boiler's life, but there is a major safety factor involved, which can save industry workers from the worst, explosions. These inspections can prevent an unplanned shutdown. With the reduction in unplanned shutdowns, the efficiency improves.

The process of manual inspection includes shutting down its operation to let it cool down. This is where the downtime increases. Once it is cool, a human enters through a utility hole and conducts a visual inspection with a torch and a handheld camera, which is highly unsafe for the workers, as the results may not be accurate. Inspecting the tubes, walls, weld joints, etc., may require scaffolding.

These inspections can be carried out using indoor UAVs equipped with specific sensors. No human entry into the boiler makes the inspection process completely safe and faster with drones. Also, it is possible to inspect the upper section of the boiler, which is difficult to inspect manually. The shutdown time decreases as drones can inspect with minimal cooling.

8.1.2 Data Acquisition and Processing

For caged drones to carry out inspections from inside the boilers, high-resolution RGB and thermal camera sensors are required to identify cracks, corrosion, hotspots, etc.

For boiler inspection using indoor UAVs, certain pre-inspection preparations must be carried out, such as shutting down the boilers to be inspected and purging their gas. Pilots need to check the entry ports and the feasibility of flying. Data collection must be carried out in vertical and circumferential scanning patterns while maintaining a constant distance from walls, either manually or in semi-autonomous flight.

Data capture during indoor inspections is done similarly to that during outdoor inspections/mapping. The videography becomes a smooth process for identification. However, when images are captured, they need to be captured in overlapping areas between the two images in front and side. Data capture must focus on welds, bends, and high-stress zones.

Key deliverables and their objectives:-

Output	**Purpose**
High Resolution RGB Images	– Visual inspection of cracks and corrosion
Thermal Images	– Identification of hotspots and insulation issues – Defect tagging
Inspection Report	– Maintenance planning and shutdown optimization

Table 8.1 – Boiler Inspection Outputs

8.2 Chimney and Stack Inspection

A tall vertical structure made up of steel, concrete, or bricks that safely vents hot exhaust gases, smoke, and pollutants from industrial processes like boilers, furnaces, etc., high into the atmosphere. This is the most common structure in any power plants, refineries, and factories. The major industries that use chimneys include thermal power plants, petrochemical refineries, cement and glass production, and chemical industries.

8.2.1 Requirement of Chimney Inspection

The structural assessment of the chimney is required to be carried out periodically, as there is a risk of crack formation, liner deterioration, corrosion, etc. These inspection reports are then generated to identify repairs. If these inspections are not conducted, there is a risk of structural failure, unplanned shutdowns, toxic gas leaks, and emission regulation violations.

When these chimneys and stacks are inspected manually, it is done using rope, scaffolding, cranes, etc. However, it is extremely risky to inspect at that height manually. Additionally, downtime is high, and the time required for manual inspection is the highest.

With indoor UAVs capable of inspecting chimneys and stacks, downtime is reduced, accuracy increases, and the risk to human life is zero, as drones can carry out safe vertical inspections with uniform coverage.

8.2.2 Data Acquisition and Processing

A high-resolution RGB camera, LED lighting, a PTZ camera with gimbals, and LIDAR can be used for chimney inspection. The PTZ cameras are Pan, Tilt, Zoom cameras with 2- or 3-axis gimbals that can move to any desired angle to capture data and check for defects from different angles. LIDARs are optional and can be used to inspect the deformation.

Chimney inspections using drones must be carried out while maintaining a steady altitude, as the drones will be flying vertically upward or downward during the inspection. Moreover, the spiral and zigzag patterns must be followed to avoid missing any structural defects.

Key Deliverables and their objectives:-

Output	Purpose
RGB Images	– Surface damage identification
LIDAR Point Cloud	– Verticality and deformation analysis
Inspection Report	– Structural safety evaluation

Table 8.2 – Chimney Inspection Outputs

8.3 Storage Tank Inspection

Storage tanks are vessels used in industries to store large volumes of oil, chemicals, water, gas, etc. There are different types of storage tanks: underground tanks, vertical tanks, horizontal tanks, external floating-roof tanks, spherical tanks, pressurized tanks, and fixed-roof tanks. A storage tank ensures safety, compliance, and efficient storage of unfinished or finished products.

Image 8.3 – Spherical Storage Tanks[90]

8.3.1 Requirement of Storage Tank Inspection

As these tanks are filled with toxic substances, it is necessary to inspect them periodically to prevent leaks and environmental hazards by ensuring their structural integrity.

When the storage tank inspection is done manually, the tanks must be completely drained and cleaned, after which gas detection is performed manually. Whereas with UAVs, minimal cleaning is required, faster inspection turnaround, and human safety at its highest.

8.3.2 Data Acquisition and Processing

For storage tank inspections, LED lighting and an RGB camera are required.

Using drones, a circular flight path along tank walls from a fixed distance needs to be flown slowly and stably, along with roof-to-floor inspection. The image overlap should be kept higher.

Key Deliverables and their objectives:-

Output	Purpose
RGB Images	– Corrosion and pitting inspection
Thermal Images	– Detection of insulation defects
Defect Report	– Asset health assessment

Table 8.3 – Storage Tank Inspection Outputs

8.4 Sewer and Drain Inspection

Sewer and drainage systems are essential underground infrastructure that collect wastewater from society and carry it away for safe disposal and treatment. Sewers and drains are different. A sewer is a sanitary sewer, the most common type, and what we most often see passing underground through societies. Storm drains are designed to collect rainwater and, in some cases, direct it to water bodies. Combined sewer and drain collect both things.

Sewers not only collect waste but also protect us by preventing diseases, protecting the environment, and mitigating floods.

8.4.1 *Requirement of Sewer Inspection*

To prevent flooding, public health issues, and structural collapse due to structural failure, sewage systems need to be inspected regularly. When done manually, a person enters a utility hole with limited maneuverability and poor visibility. Drones can provide high-quality visual data by accessing even partially flooded areas and navigating through bends and turns.

Image 8.4 – Drone Inspecting a Sewer[91]

8.4.2 *Data Acquisition and Processing*

The RGB camera and infrared camera are the key payloads for sewer inspection, along with high-intensity LED lights.

A linear flight along the sewer axis is carried out. An important point here is to maintain a low altitude. Continuous videography with a focus on blockages, cracks, and root intrusion is conducted.

Key Deliverables and their objectives:-

Output	Purpose
Video Footage	– Blockage and damage identification
Inspection Report	– Maintenance planning and clearing prioritization

Table 8.4 – Sewer and Drain Inspection Outputs

8.5 Tunnel Inspection

Tunnels are passageways built to bypass mountains or soil to provide routes and enable the transportation of various forms, viz., highways, railways, subways, etc. These obstructing bodies are natural features such as mountains, water, and land. These developments are undertaken to reduce road traffic congestion, and hence, a variety of transportation methods are emerging. However, there are certain exceptions. A drone taxi/air taxi/flying car does not need to cut through any mountains to get passengers to their destination.

8.5.1 Requirement of Tunnel Inspection

Tunnel inspection ensures structural safety and supports maintenance planning to prevent incidents. Inspecting a long tunnel is done by walking, using scaffolding, or lifts. This may lead to long traffic shutdowns or jams with one-way closures. It is difficult to inspect it manually during a set number of hours when a scheduled train is about to pass through the tunnel. In metro tunnels, there comes a requirement to block off tunnels, which is a hassle. Minimum traffic disruption occurs when drones are used instead of scaffolding. The entire tunnel can be inspected more accurately. Rapid data collection is also possible.

Image 8.5 – Drone Inspecting a Tunnel[92]

8.5.2 Data Acquisition and Processing

Tunnel inspection requires an RGB camera with LED lighting. There may be a requirement for a thermal camera for identifying water seepage and for LIDAR for deformation mapping.

The drone flights are usually conducted along the tunnel surface to inspect the side walls and ceiling.

Key Deliverables and their objectives:-

Output	Purpose
LIDAR-based 3D Model	– Tunnel deformation and clearance analysis
Ortho mosaic Tunnel Views	– Crack mapping
Inspection Report	– Structural safety evaluation

Table 8.5 – Tunnel Inspection Outputs

Points to Remember:-

- Always follow the manufacturer-provided SOP to avoid any incidents within the inspected object.
- Never experiment with flying an unprotected drone inside an object. Confined-space drones are designed for indoor inspection use only and should be considered the best choice for this purpose.
- Mission planning for the inspection using a confined-space drone should be carried out in accordance with the drone's endurance.
- Indoor applications are not limited to the above-mentioned objects or industries.

CHAPTER 09

Role of UAVs in Smart Farming

An Unmanned Aerial Vehicle is an intelligent, autonomous technology that can be used in almost every sector. Agriculture and farming are among the most important sectors that require this technology to initiate smart farming. Only a few countries use smart farming. However, in other countries, smart farming is not being practiced.

Soil testing, disease identification, crop monitoring, selecting pesticides based on crop monitoring results, and spraying pesticides and water using a spraying drone, etc., all play an important role in smart farming. All of these techniques and technologies save time, energy, labor, and costs, and greatly improve efficiency.

Agricultural land in many countries is in large quantities. Watering, spraying pesticides and fertilizers, seeding, etc., are lengthy, time-consuming tasks that require several laborers to complete. Farming needs a lot of water. In India, the agricultural sector uses 70 to 80% of the available river and groundwater. So, saving water is really important for any country. Another major issue is the use of pesticides. Pesticides are used in agriculture to protect crops. It protects crops from weeds, diseases, pests, and other threats, while preventing food losses and increasing farm efficiency. Two ways that show pesticides are used incorrectly. One, selecting the wrong pesticide to spray on the crops. Two, using more or less quantity of pesticide than required.

Unmanned Aerial Vehicles are here to address the two major problems mentioned above related to water and pesticides. Other benefits of drones include being an affordable alternative to labor and saving a lot of time. For water, there are agricultural drones for spraying. For crop monitoring, drones are used for pesticide applications. To spread the seeds or fertilizers, it can be done using agricultural UAVs.

9.1 Agricultural Sprayer Drone

An agricultural spraying UAV is used for spraying pesticides and water on agricultural farms. These drones are designed to carry a specified amount of liquid. Mostly, sprayer drones are available in sizes with liquid carrying capacities of 10 liters, 16 liters, 30 liters, and 50 liters. The dimensions and overall size of the drone vary with changes in payload capacity. For the spraying operation, other equipment and accessories are also required. These include highly durable, flexible PU pipe, a brushless water pump, high-pressure nozzles, a flow sensor, and pneumatic tube connectors.

9.1.1 Parts of a Sprayer Drone

A. **Tank** – There are different tanks available and selected as per the capacity and compatibility with the drone. Not all tanks fit all drones, so it is necessary to check which tank size and capacity will fit the drone. These tanks have one cap on top for filling with water and pesticide, and another at the bottom for the outlet.

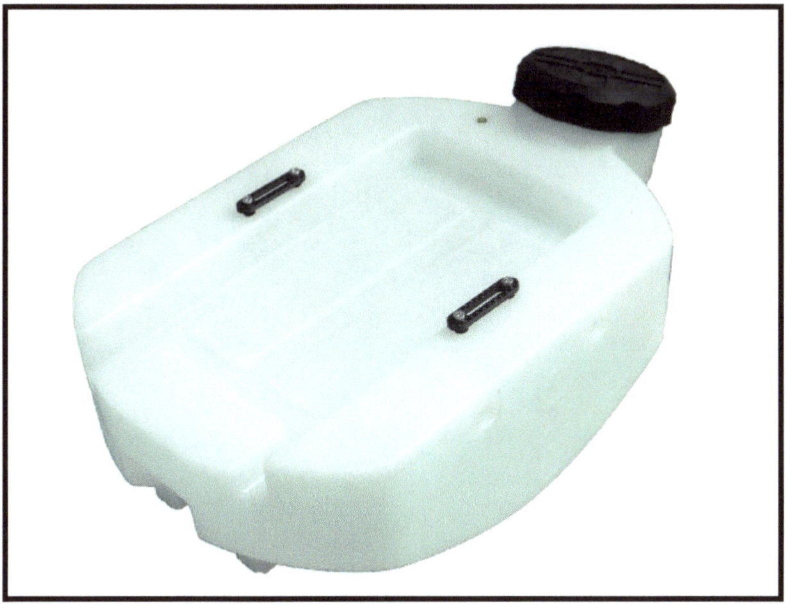

Image 9.1 – 10 Liter Agriculture Spraying Tank[93]

B. **Brushless Water Pump** - The brushless water pump is used to supply water from the tank to the nozzles. Water pumps for agriculture spraying UAVs come in different sizes. This size defines its pumping capacity and how much quantity it is compatible with. A 5-litre-capacity pump is used for an agricultural spraying drone with a 10–16-litre water capacity. An 8-liter pump is useful for a 20–30-liter water tank capacity. The 5-liter pumps have a pumping capacity of 5 liters per minute, and the 8-liter pumps have a pumping capacity of 8 liters per minute. The pump is directly connected to the power distribution board for the power supply requirement. Additionally, the pump is connected to the flight controller board. This is what helps trigger the pump to start pumping.

Image 9.2 – 5 Liter Pump[94]

C. **Flowmeter Sensor** – The flow sensor senses the flow of the water. The amount of water flowing through the pipeline to the nozzle can be controlled as per the required set limit. There is a chance that the user will not always want to spray at the 100% flow rate. Based on that, the settings can be changed and controlled using the flowmeter sensor.

Image 9.3 – Flowmeter Sensor[95]

D. **PU Pipeline** – Polyurethane pipelines are connected throughout from the flow sensor to all the nozzles using connectors and joints.

E. **Pneumatic Connectors** – Pneumatic connectors are useful as a junction, T-joint, L-bend, or extension to the polyurethane pipeline that allows the water to flow from the tank to the nozzle for spraying. These joints are reliable, reusable, and easy to connect.

Image 9.4 – Varieties of Connectors[96]

F. **Spraying Nozzles** – There are different types of nozzles available in the market that can be used for spraying water in agriculture with drones. Nozzles are mostly attached under the arms or under the BLDC motor mounting location. All these nozzles use very different spraying techniques. Most commonly used nozzles are centrifugal, hydraulic, cone, flat fan, mist/fog, etc. Also, some nozzles spray flat down 90 degrees, and some have a y-bend, so double spraying occurs. Also, there are even double–y bend nozzles available on the market that fit 4 spraying nozzles in one arm.

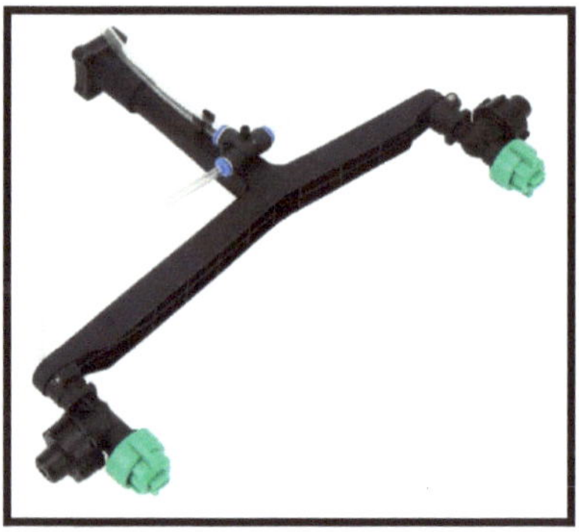
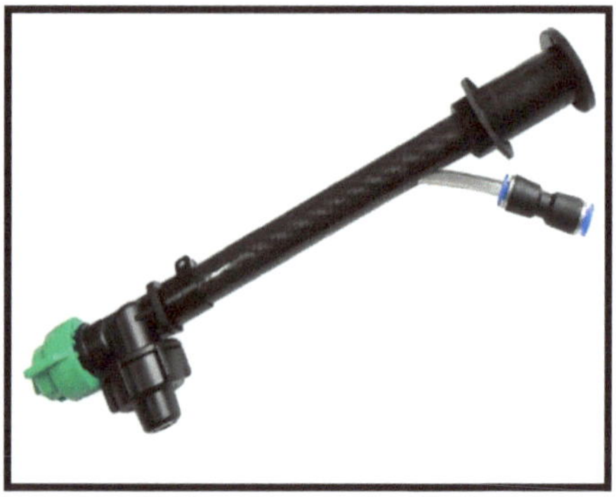

Image 9.5 – Y – Type[97] and Straight[98] Spraying Nozzles

The agricultural UAV shown below is built using the above parts and other electronics required for a UAV. This agricultural drone can carry 10 liters of liquid for spraying crops in agricultural fields. The maximum weight this drone can carry at takeoff is around 25 Kg. including 10 liters of water, batteries, and all that it needs to fly efficiently. With all these loaded on it, such a drone flies for approximately 10 to 15 minutes on a single charge and covers an area from 1 acre to more than 10 acres in a single flight, depending on the tank capacity, from 10 liters to 50 liters.

Image 9.6 – Sprayer Drone [99]

9.1.2 *Mission Planning*

Agricultural spraying missions are carried out autonomously, along with intelligent, autonomous spray control technology. Some sensors warn when the battery voltage drops to the set low-voltage level, and the tank notifies when it is empty. These UAVs are equipped with an altitude sensor and front- and rear-obstacle avoidance sensors.

Any mission-based UAV is operated using Ground Control Software (GCS), and the same applies to agricultural spraying UAVs. A mission must be planned by selecting an area within the targeted spraying field. Once the area is selected, the grid path for drone flying is automatically generated. This path can be changed as required. Drones fly around 2 to 3 meters above the crop canopy. The desired altitude above the crop can be set, and that works because sensors below the drone detect altitude relative to the surface. Once the mission and plan are ready, selection comes down to the speed of spraying over the crop and the drone's flight speed.

After every single parameter is set, the time is to crosscheck the important ones like, obstacle avoidance system is ON, Return to Home (RTH) is enabled, battery warning level is set, tank warning is enabled, etc. the drone is placed nearest to the beginning of the spraying area so that it can start spraying the earliest resulting in the low battery consumption at full load. The drone flies automatically, and there is no need to take off and land it manually. However, looking at the ground scenario of flat land and obstacles in the farm, a decision on manual takeoff and landing should be taken by the pilot.

9.1.3 *Advantages of Spraying UAV*

- ⇨ The time required to spray an area of one acre is approximately five minutes, and around ten minutes to spray a one-hectare area.
- ⇨ The total amount of water required to spray an area of one acre is reduced by more than 90% compared to the traditional manual spraying method.

- ⇨ The accuracy of spraying equally in every corner of the farm is extremely high compared to manual spraying by humans or tractors.
- ⇨ An autonomous operating system does not require more than two people (per drone) on the field, even in a large farm. This reduces the labour cost.
- ⇨ With drones, pesticides are sprayed on the crop, and they do not get lost in the soil along with water, which happens in a handheld/manual spraying technique.

9.2 Spreader Drone

In agriculture, spreading is an important task similar to spraying. This technique is used to spread fertilizers and seeds across the entire target area. Like spraying drones, agricultural drones are used to spread seeds, granular fertilizer, cattle/animal feeds, fish feed, etc.

Agricultural spraying drones have tanks that allow the water-and-pesticide mixture to be filled. Similarly, agriculture spreading tanks have a different type of mechanism that comes fitted in the tank. In case of using the same agriculture spraying drone for spreading purposes, the existing setup of the tank, that is, pipelines, pump, flow sensor, and spraying nozzles, is replaced, and the spreading tank is attached to the drone. A spreading tank has a stirring plate mechanism attached to its bottom. It rotates at the bottom of the tank once triggered for spreading. This way, it spreads the granules in the farm.

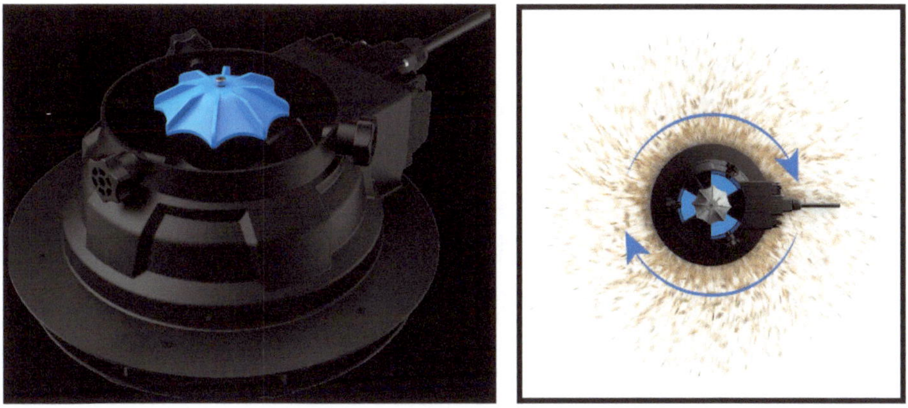

Image 9.7 – Stirring Mechanism Spreader, which Rotates 360 Degrees[100]

Image 9.8 – Drone Spreading Fertilizer Granules[101]

The mission planning and operation of the spreading drone is the same as that of a spraying drone, including warning and flight parameters.

Advantages:-

- Fast and effective spreading of granules
- Useful in multiple activities
- Saves time
- Requires less labour in the field
- Saves cost

9.3 Crop Monitoring

Crop monitoring is the process of assessing crop health, in which crops are checked for defects, diseases, weeds, etc. This real-time tracking is conducted periodically throughout the entire growth stage.

For monitoring crops, unmanned aerial vehicles equipped with a multispectral camera are used. Multispectral cameras are capable of capturing high-resolution images with multiple bands. Such sensors have RGB (visible light) capture capability, along with red band, red edge, blue, green, and near-infrared bands. The data collected across multiple bands helps analyze crop health. Additionally, it allows for the identification of vegetation and non-vegetation (and unhealthy crops).

A. Crop Health Monitoring

Crop health monitoring is the process of monitoring crop health and growth stages to identify diseases and deficiencies early. This early-stage detection helps in overcoming the issues/diseases the crop is growing under.

Payload Required – To perform this process, data for the targeted farm area must be collected. This data is then processed to yield the best results, allowing the user to perform manual activities on the crop. For this purpose, a visible light sensor/camera will not work. This requires a multispectral camera sensor that captures images in multiple bands – red, red edge, green, blue band, along with an RGB image. This entire process can be done quickly and easily with the help of a UAV because manually identifying the disease in the crop is very difficult and time-consuming. In a single flight, drones can capture images of around 10 hectares.

Mission Planning – A mission must be planned using ground control station software, as in other applications. When using multispectral sensors, there are procedures for installation and calibration that must be followed. For crop monitoring applications, multirotor UAVs are preferred. Such a mission can be planned and executed autonomously.

Data Processing

Data capture for crop analysis can be done with grid missions, where the drone follows a predefined path and captures images in 5 or more bands according to the multispectral sensor specifications. Data from all bands can be viewed and retrieved from the same SD card, which contains all data collected during the flight. One image will have 4 or 5 similar copies from different bands. This data is then processed for the desired output.

Key deliverables and their objective:-

Output	Purpose
Ortho mosaic	– Crop boundary monitoring – Growth stage monitoring
Vegetation Index Maps	– Stress and variability analysis

Table 9.1 – Crop Health Monitoring Outputs

B. NDVI

The Normalized Difference Vegetation Index is a calculation-based index derived entirely from multi-band images because healthy vegetation contains chlorophyll that reflects more NIR (Near Infrared) and green light than other wavelengths. Whereas it absorbs more red and blue light. The formula for NDVI calculation is given below:-

The calculation of NDVI typically yields a value between -1 and +1. A negative value indicates a water body or barren soil, whereas a value close to +1 indicates a higher likelihood of densely green leaves and healthy vegetation. When this NDVI value is close to zero, it indicates that there are no green leaves and may indicate an urban area. The higher the NDVI value, the better/healthier the vegetation.

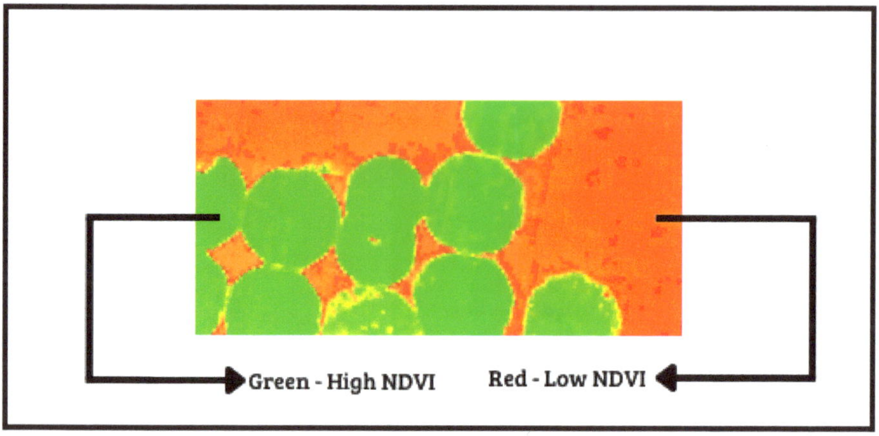

Image 9.9 – NDVI Data[102]

NDVI data is very useful for precision farming. This data can be collected and analyzed by central government organizations, state government remote sensing centers, privately owned farm lands, farmers, etc. NDVI data can be generated from satellite data; however, for high spatial resolution and precision, drone-captured data is recommended.

Flights planned with a multispectral camera may require an overlap of around 70% or more for precision monitoring. The flying altitude is generally set around 80 meters.

Key Deliverables and their objectives:-

Output	Purpose
NDVI Map	– Chlorophyll and biomass estimation
Zonal Statistics	– Field – wise crop health comparison

Table 9.2 – NDVI Data Outputs

Points to Remember

- Select the takeoff location nearest to the spray starting point for the best efficiency.
- Agricultural spraying and spreading UAVs are heavy equipment, so for safety, it is recommended to use only an RTK-enabled system.
- Multispectral sensors need to be calibrated and may have certain operating temperature limitations.

CHAPTER 10

UAVs for Professional and Entertainment Purpose

Entertainment of all kinds, whether it is video games, sports, film and television, news coverage, etc., has always embraced technological upgradation to enhance how it is captured and reduce the production timeline. In the entertainment industry, including photographers, videographers, and filmmakers, traditional methods with bulky equipment have long been relied on for aerial or topographic shooting, where helicopters or cranes are used to capture the shot.

Coming to games and sports, so far, society is aware of video games that used to be played on our television, with the game stored on a cassette attached to the gaming device, which was then connected to the TV. And now, there are gaming apps with a real feel. Also, PlayStation and VR games are popular among gamers, letting you play games practically all alone in your room. No human required physical labor, and no injury to anyone. The setup for the sports used to be something like this: a cameraman sitting in a chair with his camera on a rail track inside the circle of an athletic ground, capturing the entire race on video. Even today, cricket and football stadiums have suspended-cable camera systems that cover the field horizontally and vertically. However, the technology has advanced, and these days there are drone games that combine gaming and sports in a single event.

The following are a few use cases for the entertainment, media, and marketing industry.

1. Drone as a business and profession
2. Drone for leisure flying and competitions
3. Gaming
4. Professional use case – marketing, media, and advertising

10.1 Drone as a Business and Profession

Photography and videography serve a wide range of audiences and users worldwide, ranging from mobile photography to waterproof action camera photography and from film shooting to sports photo/video shooting. Before the digital camera came to market, there was thin-film camera technology, in which the limit on image capture was set by film reels containing around 36 frames. Meaning, one can capture only 36 photos with that reel. There was no way to tell from the captured image that it was blurred or that someone's eyes were closed. With the emergence of digital cameras, the drawbacks of limited image capture were eliminated, as was the need to playback captured images. The use cases of image/videography are given below:-

a. Leisure photo and videography, including travel vlogs
b. Film shooting
c. Sports photography and videography
d. Professional and event shooting

Cameras have now taken one step beyond handheld and tripod-mounted photography by being attached to a UAV. With drones, mobility and shooting have reached a whole different level. New scopes of capturing images have emerged. Compared to being limited to the ground, aerial imagery can capture photos from any height and at any angle. Capturing and zooming in on hard-to-reach areas is now possible with drones.

The never-felt, never-seen bird's-eye view is something everyone must capture to relive the moment after years of imagination.

Drones have opened up possibilities for image and video capture to such an extent that all four of the above-mentioned use cases can benefit from upgraded technology in their fields, making image capture and distribution, along with live view, faster than before. Film shooting gets easier with drone cinematography and the required cameras, which can shoot multiple shots as needed, with the added advantage of mobility.

With drones, sports photographers do not need to run alongside an athlete to cover the performance and can easily capture it with full control in the user's hands, or autonomously.

Overall, drones for photography and videography are available in various sizes to meet users' requirements.

10.2 Drones for Leisure Flying and Competitions

Things look different when viewed from above than when viewed from the ground. In drone photography, the bird's-eye view is considered breathtaking, especially when there is a beautiful natural landscape all around. It is always an overwhelming feeling to see everything from the sky while flying a drone, just as it is when you look across the streets while driving your vehicle.

The First Person View (FPV) is a technology found on certain drones, in which the camera is mounted at the front of the drone.

Image 10.1 – An FPV Racing Drone[103]

The image above depicts an FPV racing drone. The dimensions of this drone vary with the propeller size. In the image above, there is a camera at the front. These cameras can be analog or digital, depending on the drone specifications and user requirements. Wirelessly connected to these cameras, the video transmitter (VTX) and video receiver (VRX) pair communicate and share the drone's live video with the user. These receivers, on which the user gets the live video feed, can be a static display system or goggles for an analog camera system. For digital camera systems, goggles are mandatory with the inbuilt or compatible external receiver.

Image 10.2 – Analog FPV Camera[104]

The above image shows an analog FPV camera that is attached to the front of the drone, providing us with a bird's-eye view. The output, live video feed, is displayed on the analog goggles paired with the drone. The image is shown below. These goggles are similar to VR glasses and must be worn as a headset. There are setting buttons that let the user set the channel to match the analog video transmitter circuit's channel value to receive output from the desired device.

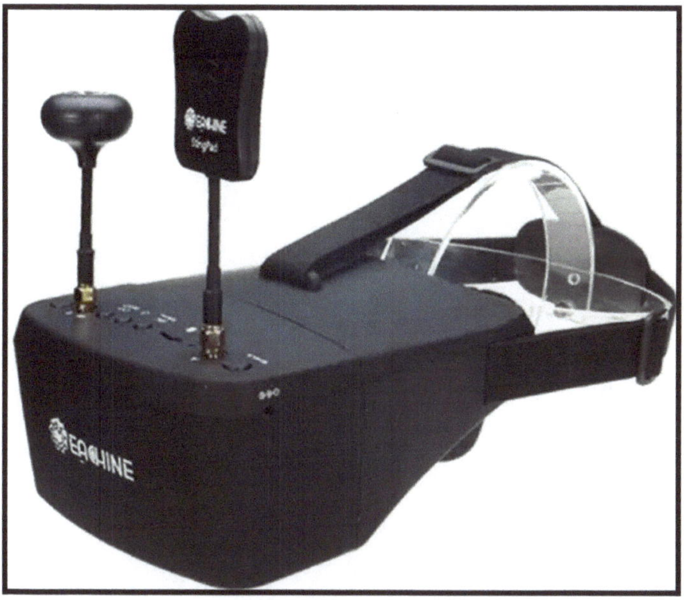

Image 10.3 – Analog FPV Goggles[105]

There is a difference between analog and digital imaging. Digital images captured are always clear and better than analog images. Similarly, there is a significant effect of a digital live data feed captured by a digital camera and transmitted via a digital video transmitter to digital goggles.

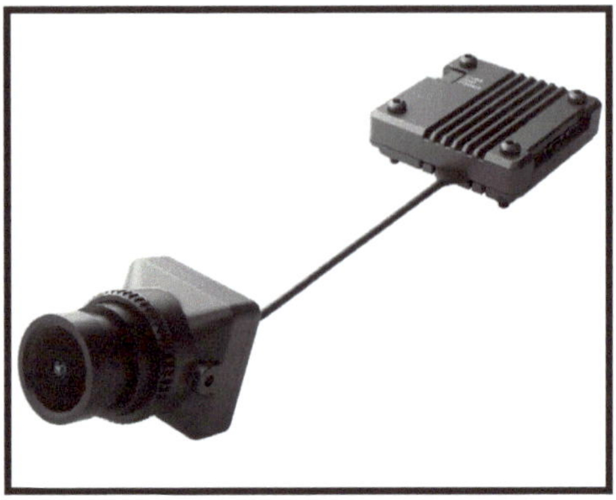

Image 10.4 – Digital FPV Camera[106]

The image below shows a digital FPV goggle used to view the live data feed from a drone equipped with a digital FPV camera. The display quality and range of such digital devices are better than those of analog devices, and they are cheaper than their digital counterparts. The main concern for any digital unit buyer is compatibility. Not all goggles are compatible with all receivers. Specific goggles have a specific VTX and VRX set to communicate and transmit data. So, after buying one pair of goggles, it becomes difficult to switch to another brand or device quickly. FPV goggles are costly if they must be purchased for each drone.

Image 10.5 – Digital FPV Goggles[107]

The complete set of analog versions is cheaper and useful for all users learning to fly drones in FPV mode. The analog set's range is limited, and the display may show distorted lines when the drone goes a little too far or passes behind buildings. This entire set is used for hobby and Vlog flying.

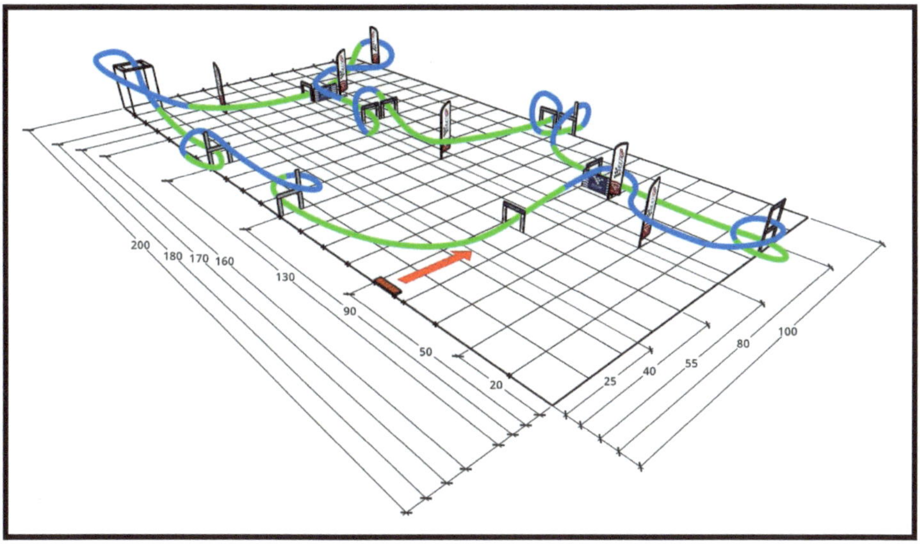

Image 10.6 – Drone Racing Track[108]

Over the last few years, drone racing has become one of the most sensational experiences among FPV drone fliers. The fliers compete in competitions such as the Drone Racing League, where drones race on a provided track, as shown in the image above. This track includes a variety of hurdles and pathways, and an organizer provides the terms and conditions for flying, as well as drone specifications. This is one of the emerging fields for hobbyists and leisure fliers. This hobby will surely increase your adrenaline.

10.3 Gaming with Drones

Football is the God of all the games, but very few get the chance of playing on the field. Drones have made it possible for everyone to play soccer. In simple terms, this is a game played with flying drones. And there is a championship that is organized at an international level competition. The Federation of International Drone Soccer Association (FIDA) is the body that has set the standards and rules for playing this game, so that everyone in the world who plays it can participate in the international tournament.

The regulations include the following:-

a. Standard layout of the arena – surface leveling, marking lines, dimensions in length, width, and height, skyfence wall covering, dimensions of the goal post, goal post location, material standards of the goal structure, and advertisement standards.
b. Drone Ball – Dimensions and specifications of the drone ball, advertisement standards, officiality of the drone ball, LED marking on the drone ball for the team identification, color coding of the team players and color coding of the team members (position-wise – striker, guide, libero, sweeper, keeper), and usage of frequency for drone operation.
c. Team Members – Total players in the team, substitutions, substitution procedure, striker substitution during the play, violations, and penalties.
d. Player's Equipment – Basic equipment (attire, drone ball, controller, battery), additional equipment (FPV goggles, extra drone ball, other accessories), prohibited equipment, violation and penalty regulations, advertising standards, etc.
e. Head Referee – Authority of the head referee, powers and duties, position of the head referee, decisions of the head referee, responsibilities of the head referee, head referee qualification, signals standards, etc.
f. Assistant Referee – Classification and role, powers and duties, qualification, assistant referee signal standards, etc.

g. Start and End of the Game – Number of sets and play time, game preparation, start and end of a set, repairs and suspension, start of the next set, abandonment of the game

h. Offense and Defense – Score, offside, defense

i. Penalty Kick – Penalty award, offset of penalty, penalty kick method and procedure

j. Winning Decisions – Winning team, draw, penalty shootout

k. Fouls and Misconduct – Types of fouls, warning, forfeit of the set, forfeit of the game

A few example images are shown below to illustrate the drone soccer game.

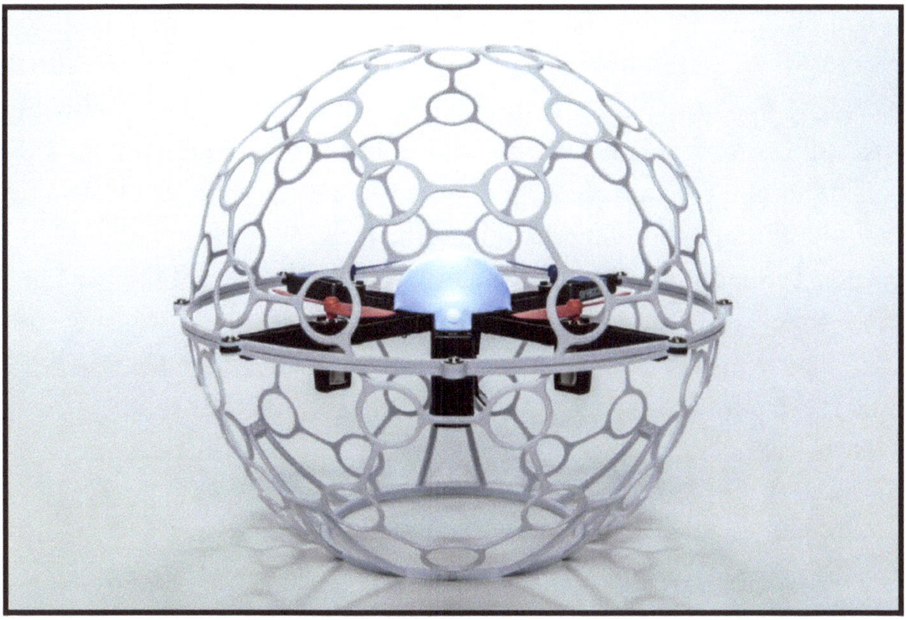

Image 10.7 – A Replica of a Drone Ball[109]

The above-mentioned image is a replica of a drone ball. The drone, in most cases a quadcopter, is secured inside a cage to prevent it from colliding with any other drone, goalpost, or the arena boundary.

The diameter of this ball must be around 400mm, and the weight of the complete drone ball must be less than 1.1 kg.

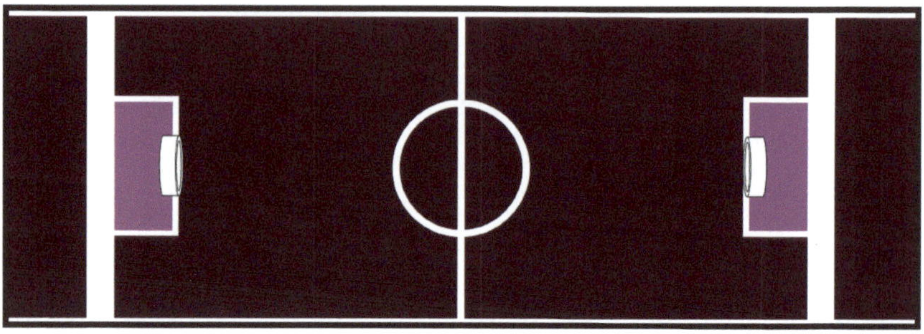

Image 10.8 – Drone Soccer Arena Layout[110]

The drone soccer arena has a defined layout of the entire stadium, including dimensions. The playing arena must be 10 to 20 meters long and 5 to 10 meters wide. That is, the ratio of length to width must be 2:1. The height of the fence is kept at 4 to 5 meters. The goal post is placed 1.5 meters from the boundary line and should have an inner diameter of 60cm and an outer diameter of 100cm. The goal's thickness should be 20 centimeters. Whereas, the location of the goal post in the center is fixed at a height of 3 meters from the bottom of the sky fence. The passage for the players, or the drone pilots, to control the drone and play is provided at both ends of the long sides. The passage width is 2 meters, and the length is equal to the width of the arena.

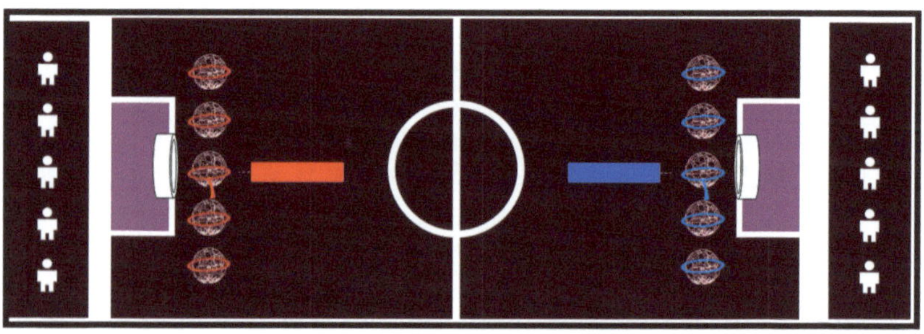

Image 10.9 – Players and Team Position[111]

The above image shows the team's position inside the arena. The teams are identified with color coding. Two LED colors are declared – red and blue for a particular team. In one game, there are three sets, and five minutes are provided for each set. There are other set regulations too, related to flying and other practical aspects, which can be understood by the pilot and the team on the real game field.

10.4 Professional Use Case – Marketing, Media, and Advertising

Product marketing is crucial, as everything depends on the product's visibility. The motto of any product marketing is to get people to know about it and then think about buying it. Creating awareness among the target society is the most important. Then come the other important parameters, like building trust in society for the company and the product, comparing quality and pricing, promotions, etc.

Marketing on regular billboards and ad-boards is still there, and on roads, there's a chance of overlooking it or missing what's on it. Drones, on the other hand, catch the eye with a unique way of publicity, never seen before. It helps in engaging the audience by capturing images and videos from unique angles.

A. Branding

A drone light show is one of the best ways to promote an organization and its product branding. In this, a particular brand/company name, logo, and its total years of presence can all be showcased using drone light shows. This automatically attracts the audience, and people become aware of the brand in their area. This marketing strategy definitely boosts the brand awareness in society. However, this approach is a temporary marketing strategy, but the promotions are not limited to a specific audience and may even raise awareness of the product/company in international markets.

B. Virtual Tours

Real estate firms can use drones not only to track the construction timeline but also to demonstrate the properties. Strategic drone videography can be shot and edited professionally, so that when viewed, it provides a virtual tour of the entire property through the goggles. Additionally, a walk-through can be provided, saving the buyer time from visiting properties multiple times.

Another marketing formula is banner towing, in which a fixed-wing UAV carries a banner bearing a brand name. This and certain other fancy solutions that use drones for marketing and advertising are catchy and more quickly impress the audience due to the company's unique branding.

In addition to the above-mentioned professional use cases, drones can also be used for warehouse inventory management.

Points to Remember:-

- The caged UAVs developed for indoor inspection and drone soccer are entirely different from each other.
- Racing drones and drone soccer are meant for entertainment, gaming, and sports purposes. However, both of these require practice flying, both solo and in teams.
- A small mistake during the operation of an FPV racing drone may lead to a crash or the drone's complete loss.

PART IV

FUTURE TECHNOLOGIES AND EMERGING APPLICATIONS

CHAPTER 11

Role of UAVs in Modern Defence and Military Applications

The defense sector, in any country, is much more important, as it is responsible for protecting the country from external aggression, threats, terrorism, etc. Defense organizations protect the nation and its territorial integrity by securing its borders, responding to external threats, and maintaining internal security.

The primary objective also includes border protection, where the forces monitor and secure the land, air, and maritime borders. Additionally, these forces often assist in rescue operations and emergency infrastructure during disasters such as earthquakes and floods. With a focus on technological development, the defense sector plays an important role in advancing technologies in areas such as aeronautics, robotics, communications systems, and satellite technologies.

11.1 UAVs in the Defense Sector

Drone Technology plays a significant role in the Department of Defense during modern defense operations by the Army, Navy, Air Force, and Coast Guard. UAVs are developed with a focused interest in their use case, which is why they are designed and built precisely for a particular application. Depending on the requirements, drones are designed and developed or selected for purchase, which can be: longer flight time, high altitude operations and surveillance, dangerous missions like border surveillance, and suicide missions on the targets, etc. A few major roles of the UAVs in the defense sector are described here.

UAVs are capable of providing real-time imagery and video feeds to the pilot and the department, enabling the team to monitor activities across the border, enemy positions, wartime surveillance, infiltration attempts, etc. This is possible with the help of ISR – Intelligence, Surveillance, and Reconnaissance. Target detection and tracking is another application for which UAVs are developed. This includes identifying enemy vehicles or persons and continuously monitoring them.

Drones are also playing the role of modern fighter jets, without any pilot on board, by carrying missiles and guided munitions to strike the targets precisely.

11.2 Generations of Warfare in the Defense Sector

The defense industry advances gradually and in terms of generations. These are the generations of warfare[112] which mostly reflect their current stages:-

- Level of autonomy
- Human involvement
- Network integration
- Information dominance
- Decision – making speed

The generations of warfare are different from the UAV technology generation, as the warfare generations show us how wars were fought. Whereas UAV technology generation describes the evolution of drone technology.

A. First Generation Warfare

The first generation of defensive warfare lasted until the mid-1800s. The core focus of defense was on infantry-based combat vehicles and not on unmanned systems, at that time. This was the reason why there was almost no possibility of surveillance. Mass manpower and basic weapons were used for defense.

B. Second Generation Warfare

Second-generation warfare began in the 1860s and lasted until 1936, when mechanization took place, and heavy firepower became the primary focus. This includes weapons such as tanks, artillery, and aircraft. This was the period of human-piloted systems, mostly used for early-stage reconnaissance. There was no use of UAVs till the Second World War.

C. Third Generation Warfare

The Cold War was part of the third generation of warfare, during which satellites were available for reconnaissance. Precision-guided weapons were adopted during this period, which includes fighter aircraft and missiles, with the aim of surprising the enemy with speed. Additionally, this was the phase when Remotely Piloted Vehicles were under early-stage experiments. This period from WW2 to 1989-90 is considered as 3rd generation of warfare.

D. Fourth Generation Warfare

Currently, the defense sector is undergoing fourth-generation warfare, from the 1990s to the present, during which networking technology has advanced and network-centric operations and ISR have been adopted. This was also the period of the rise of surveillance and strike drones.

E. Fifth Generation Warfare

The fifth generation is the future of the defense sector, with a core focus on cyber warfare, electronic warfare, AI-assisted decision-making, swarm-based warfare, etc. This generation has emerged since 2014-15.

11.3 Evolution of UAV Technology

The evolution of UAV technology has occurred gradually through several technological phases. From the above generations of defense warfare, it is clear that the actual use of UAVs in military applications began after 1950. However, UAV technology has been under development and in the experimental stage since the 1900s. The following phases describe the evolution and advancement in communication technology, control system improvements, sensor innovations, changes in battery technology, and levels of autonomy, which resulted in the overall evolution of UAV technology.

A. Early Experimental UAVs

This is the earliest and longest stage, with the initial phase of attempts at the design and development of UAVs for military applications. The core focus of such UAVs was primarily on experiments for target practice and initial-level reconnaissance. These drones were operated with pre-programmed missions or manually via radio/remote control, with no autonomous flight capability available on board.

The early stages of UAVs were limited. The biggest limitation was the operator's lack of access to a real-time data transmission facility. The range and endurance of these drones were once very limited. The UAVs of initial duration were unreliable for the desired operations.

Image 11.1 – Ryan Model AQM - 34[113]

During this earlier phase, the Ryan Model 147 Lightning Bug was considered one of the most useful UAVs developed by Ryan Aeronautical in the 1960s and became part of the US Air Force during the Vietnam War for reconnaissance.

B. Development of Tactical UAVs

The 1980s to 1990s are considered the period when UAV technology began to evolve into a reliable tactical system for battlefield reconnaissance, border security, surveillance, ISR, and more. During this period, tactical UAVs were equipped with an advanced radio communication system, a live data/video transmission system, and digital imaging sensors, along with improved stabilization and navigation systems.

With such technological advancements, it was possible to observe the battlefield in real time, which significantly reduced the risk of manual surveillance flights. Even after the technology upgradation, there were certain limitations. The communication range was short, and the entire operation still depended on the ground control operator.

Image 11.2 – IAI Scout[114]

The above image is from Israel Aircraft Industries' IAI Scout UAV, developed in the 1970s and first operated in 1981 during the combat missions by the South African Defense Force against Angola[115].

C. Armed UAVs

The capabilities of UAVs have further advanced, leading defense forces to use them for satellite-based long-endurance missions. This was during the period of 2000 to 2015. Also, there was a significant improvement in flight time, allowing the drones to fly for more than 24 hours. The payload systems were further improved with high-resolution EOIR sensors and an upgraded carrying-and-dropping mechanism for the weapons. Additionally, these UAVs were integrated with the latest navigation and autopilot systems. These UAVs were able to conduct precision strikes and continuously surveil the targets.

Image 11.3 – General Atomics MQ9 Reaper[116]

In 2007, USA-based General Atomics introduced the MQ 9 Reaper UCAV. Also known as Predator B, it is a MALE UAV capable of carrying approximately 1300 kg of payload and flying up to 50000 feet for more than 24 hours. These drones are typically armed with AGM-114 Hellfire missiles, GBU-12 Paveway II, and GBU-38 JDAMs[117].

D. Autonomous UAVs

This phase is ongoing when UAVs have the capability for autonomous navigation and obstacle avoidance. The core focus of the defense sector in this phase, from 2014-15 to the present, has been to improve maritime and border security through UAVs and mission flexibility. Additionally, these UAVs were multi-role mission-capable and had better integration with military networks.

Image 11.4 – Elbit Hermes 900[118]

In 2010, Elbit Systems introduced the Hermes 900 UAV, which Israel first used in July 2014 during Operation Protective Edge[119]. Hermes 900 is a MALE UAV capable of flying for around 30 hours at an altitude of 30000 feet and carrying a payload of around 400 kg. This is one of the most accepted UAVs, with over 20 nations using it for ISR, maritime patrol, and precision strikes.

E. UAVs with Collaborative Operations

This phase shows how drone technology will evolve in the future. The next generation will have a focus on Swarm-based surveillance and attacks. All of this will happen through the coordinated operations of multiple UAVs, AI, inter-drone communication and cooperation, etc. This technology will be beneficial for high operational scalability, tricking enemy defense systems, and increased redundancy. This phase will be the future; however, defense research programs across the World are actively developing and experimenting with their capabilities for full-fledged use.

11.4 UAVs from the Future

UAVs are now evolving faster compared to their initial stages. As discussed above, if this evolution is considered in phases, the time taken for the technology to evolve from the first phase to the second was 50 or maybe 60 years, whereas it was just 15 years for the UAV technology to evolve from phase 3 to phase 4. The fifth phase covers UAVs currently under development and expected to be adopted in the very near future. The possible upcoming technologies in the early stages of development are listed below. The first appearances of these technologies are evident today, but they will be widely adopted and become strategically important in the upcoming phase of UAV technology's evolution.

A. Loitering Munitions/kamikaze Drones

Loitering munitions, or kamikaze drones, are designed and developed to hit targets precisely. These suicide drones will be combining surveillance along with explosives as a single-use weapons system. First, these drones loiter over the target area and then use the integrated explosives to destroy themselves along with the target. That is why they are also known as suicide drones.

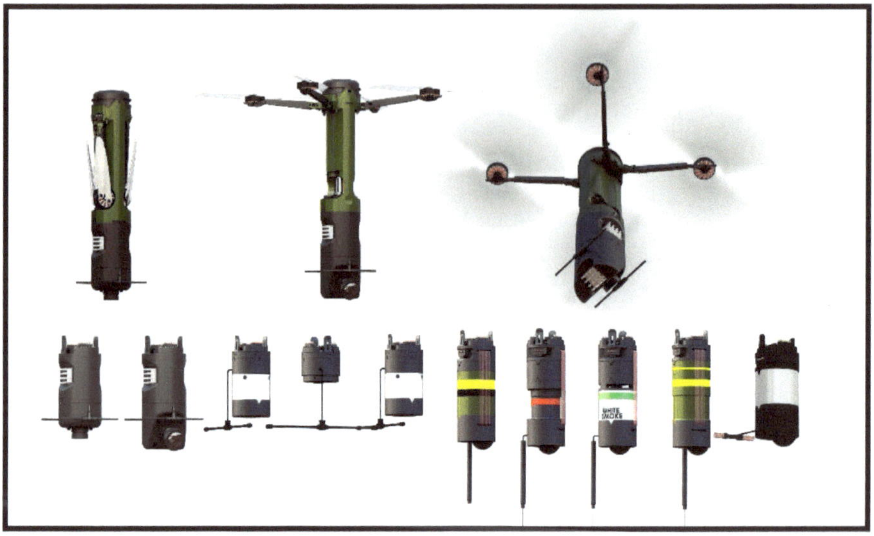

Image 11.5 – Drone 40[120]

The image above is of the Drone 40 from the Australian company Defendtex. In this image, the upper portion shows the drone in a foldable, deployable configuration, while the payload options are shown in the bottom portion. This loitering UAS can fly for approximately 1 hour and has a range of up to 35 km with an MTOW of 300 grams. This drone is designed with a modular bay that supports multiple payloads, allowing selection and attachment before flight. This includes ISR imaging, ISR thermal imaging, training effector, frag, smoke, EFP detonator, IR streetlight, EO streetlight, etc.[121]

With this technology, the defense sector's core focus is to strike precisely at high-value targets and conduct anti-armor operations to destroy weapons, radar systems, and air defense systems. Adoption of this technology is beneficial to a country's defense sector, mainly because of its low cost and rapid manufacturing capability compared to missiles, which are essential, especially during wartime.

B. Swarm of Drones

Multiple UAVs coordinating to accomplish a mission form a swarm of drones. In this, a single drone does not perform the operation. Instead, hundreds of drones coordinate to complete the assigned tasks. These drones are equipped with inter-drone communication, decision-making, cooperative target engagement, etc., with a core focus on trapping operations and wide-area surveillance.

A swarm of drones can saturate enemy defense systems, and some may not be intercepted. Even if a swarm of drones is detected, it is difficult to destroy all of them in one go, so the possibility of redundancy is higher. In terms of surveillance, these drones can cover a large area in a single go.

C. Navigation Resilient UAVs

Navigation-resilient drones are integrated with the latest technology, including programmed hardware boards that can provide alternative navigation methods. This technology is evolving and under progress with the view of mitigating the risk of GPS spoofing. Such incidents have occurred many times in the past, and there is a high risk of GPS spoofing even in the future, which can affect the operation of autonomous UAVs. During such a period, an anti-GPS spoofing drone can operate in GPS–denied environments. However, this system increases the overall product cost.

D. Cargo Delivery Drones

UAV dropping its payload, whether they are dropping weapons, guided missiles, packages, etc., does exist. A very good example of the same has already been shown in the above topic with the Elbit Hermes UAV and the MQ-9 Reaper UAV. They both are capable of carrying and dropping heavy weapons on enemy targets. However, the transport of weapons, medical supplies, etc., using drones is still in process.

Image 11.6 – Griff 60[122]

The above image is of Griff 60, developed by a Norway-based company, Griff Aviation. This Octa quad drone is capable of carrying a 60 kg payload and flying for around 30 minutes. Such a system is indeed required to fulfil the heavy delivery requirements of defense, as it reduces the overall timeline for defense supplies, which is very important for battlefield logistics and supply in remote or dangerous areas. These drones are built such that it can carry heavy payloads. Drones that can take off and land vertically are preferred, as they enable precise execution of such a mission when equipped with autonomous delivery capability.

E. Fiber Optics Controlled Drones

Like GPS spoofing, there is also a possibility of frequency jamming, which targets the radio-controlled communication frequency of the drones. To protect wireless drones from jamming, wired communication is used. A physical fiber optic cable is used to control the drones instead of a radio controller. As there is no wireless communication, it cannot be jammed. These drones are capable of operating in electronic warfare environments and can precisely strike targets in heavily jammed battlefields. However, these drones have certain limitations. The operating range is limited due to the use of a physical cable with a length of approximately 1 km to 15 km. Other challenges include cable management and a restricted ability to move compared to wireless drones.

F. Loyal Wingman Drones

Defense UAVs came after fighter aircraft, and now defense sectors require these to team up. Loyal Wingman Drones are also known as escort UAVs, which team up with fighter aircraft and operate together, with the UAV assisting pilots with reconnaissance, electronic warfare, and weapons carriage, among other tasks. This expands the operational capabilities of fighter aircraft while reducing risk to pilots.

G. Stealth UAVs

Stealth UAVs are designed to quietly enter enemy territory without being detected by enemy radar or defense systems. The stealth capability increases survivability during covert missions by reducing the risk of detection.

Points to Remember:-

- The demand for UAVs and UAV technology in the defense sector is growing.
- Majority of Defense UAVs are developed as BVLOS capable systems.
- The requirement for drones in the defense sector is based on the variety of applications, as multiple drones are needed for different missions.
- The production of defense UAVs should be designed to enable rapid, large-scale development during the war period.

CHAPTER

12

Artificial Intelligence in UAV Systems

Unmanned Aerial Vehicles, as the name suggests, are vehicles operated without a pilot or operator onboard, and are therefore equipped with sensors, cameras, communication systems, processors/computers, etc. So far, whatever applications we have gone through, UAVs are mostly operated "autonomously" using a pre-planned or pre-programmed mapping route of the targeted area, or they fly when a pilot operates them, "manually" controlled. These drones and UAVs have been available in the market for a few years. Every part of these UAVs changes with time. It could be due to smart manufacturing methods or another reason, but the ecosystem is moving ahead. To move a little faster and smarter, the adoption of AI in UAVs is a must.

With artificial intelligence, we can add perception, decision-making, and learning capabilities so that it can operate autonomously in various environments intelligently. Using AI, drones can understand what they observe. They can interpret visuals, identify objects, detect anomalies, and make real-time navigation decisions such as following an object with live tracking.

The following components play an important role in integrating AI in UAVs.

A. Sensors – These can be high-resolution cameras, LIDAR, Depth sensors, thermal cameras, GPS, etc. These sensors provide raw data for the AI-based processing

B. Flight Controller Integration – These provide the low-level control and actuation. The AI computer sends commands to the flight controller based on the decision-making capability it has learnt.

C. AI Algorithms and Models (Software) – To fulfill the requirement of decision making, navigation-guidance-control, and perception, computer vision, Deep Learning, SLAM, and Predictive Analytics Models are used.

D. Edge AI Computer (Hardware Platform) – High-performance processors that run the AI Algorithms and Models, such as SLAM, Target Tracking, Path Re-planning, and decision-making logic.

The above parameters are so robust that they can be used in more than 100 applications. There are many applications covered in this entire book. Here is a list of applications where the above components of AI integration can be implemented to increase the efficiency of any UAV, simplifying tasks and automating data processing:-

- ⇨ Obstacle detection and real-time avoidance
- ⇨ GPS denied navigation using SLAM technology
- ⇨ Landing precisely on a moving platform
- ⇨ SWARM coordination and collision-free multiple UAV operation
- ⇨ Object detection and target tracking
- ⇨ Automated Photogrammetry and 3D reconstruction
- ⇨ Crack detection
- ⇨ Hotspot and damage detection
- ⇨ Defect identification
- ⇨ Object counting
- ⇨ Intruder detection and breach alert
- ⇨ Crowd density estimation
- ⇨ Dynamic route optimization
- ⇨ Traffic monitoring and congestion prediction
- ⇨ Victim detection under debris
- ⇨ Flood and water level mapping
- ⇨ Intelligent data filtering
- ⇨ Automatic report generation from flight data

The above list is extremely short, and one can add unlimited use cases to it. A few of the applications are explained below to provide an overview of the impact of using drones along with AI:-

1. Object Counting
2. Target detection and tracking
3. UAV Swarming
4. UAV Lightshow

The identical or interrelated applications are selected so they can be covered more effectively.

12.1 Object Counting

The object counting process involves detecting objects, such as people, vehicles, animals, trees, infrastructure, and farms. It takes place to determine the specific number of objects present in a particular targeted area.

Although it is not a current problem, it will definitely be addressed with the help of technology. However, the traditional method used to follow the few typical, not-so-precise, and still time-consuming methods, such as manual surveying. It is possible to count static, non-living objects at fixed positions, but it is way too difficult to count living, moving objects manually. Another way, which later came up, was the use of CCTV cameras, which can detect and count objects only if they pass within the CCTV's field of view.

Drone + AI as a Solution – UAVs, when combined with AI-based vision systems, can provide a highly effective and efficient solution for applications like counting. This happens because of the following advantages:-

a. Fastest data acquisition of a larger area
b. Deep Learning models can detect and identify the objects in each frame

c. Objects like – vehicles, trees, structures, etc. get automatically counted by the system

d. UAVs can periodically scan the same area using the same flight plan for analysis at regular intervals.

e. Access to difficult areas remains open. For example, vehicle scanning on a congested highway, people counting during a natural disaster, animals in the forest/sanctuary, etc.

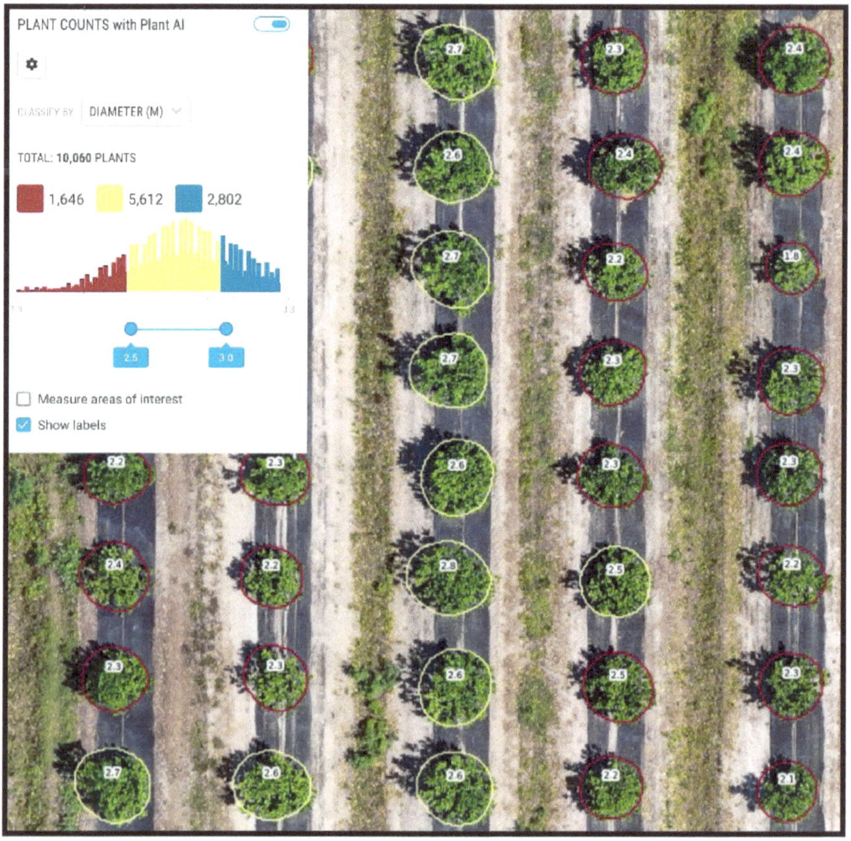

Image 12.1 – Object Counting Using Drone and AI[123]

The image above is from the plant counting solution provider, which provides an automatic plant counting system. This is a software-based system that automatically counts the trees in a selected or targeted area, compared to manually clicking each tree and counting it.

Few Applications of Object Counting are:-

a. Crowd counting – event management and safety
b. Inventory counting – in ports, mining stockpiles
c. Wildlife population estimation
d. Vehicle counting for traffic analysis

UAVs with AI can make things faster by using Edge AI and counting algorithms, which not only help with object detection but also with counting objects in real time, with high resolution, high accuracy, efficiency, and automation, reducing the effort and time required for manual data analysis.

12.2 Target Detection and Tracking

Target detection stands for the identification of the specific target, which can be a person, a vehicle, an animal, or any other object of interest, and to understand its presence as well as location. In contrast, target tracking can follow the target's movement in real time. So, detection is about checking "What is there?" and "Where is it?" Whereas, tracking is to identify "Where is it now?" and "Where is it going?"

The previously explained AI-based application for object counting provides the total count and does not track object identity. The focus of object counting is *quantity,* whereas detection and tracking focus on *identity and motion.*

It all started with human-based surveillance, where patrol teams used to visually observe the target using leaked information and tips, then chase them. Again, the CCTV cameras with multi-axis vision were used. These PTZ cameras can detect and identify objects from different angles, but they cannot move, so coverage remains limited to the fixed infrastructure.

UAVs can follow the target from above the ground, and in any direction, repositioning themselves dynamically to maintain the line of sight. Not just that, it can automatically adjust the flying speed, altitude, and heading without any human input. All of these are implementable with AI-based visual tracking, where Deep Learning helps detect the target in real time. The target identity is maintained across different frames using a tracking algorithm, and future movement is estimated with the help of predictive models.

Few applications of target detection and tracking:-

a. Wildlife monitoring
b. Studying wildlife migration
c. Border and Coastal surveillance
d. Search and Rescue
e. Traffic Violation monitoring

Target detection and tracking involve identifying specific objects and continuously monitoring their movement over time, while maintaining their identity, position, and trajectory. AI-enabled UAVs combine mobility and real-time intelligence in order to detect targets and track them accurately and cost-effectively.

12.3 UAV Swarming

The multiple agents operate together by following a set of standard local rules, which results in coordinated, collective behavior to achieve complex objectives through cooperation. This is a nature-inspired system, such as flocks of birds flying in different shapes, schools of fish, and colonies of ants, which move cooperatively.

When the same coordinated operation using multiple UAVs is executed, it becomes a swarm of Drones or UAV Swarming. The drones communicate and collaborate to perform a mission. Each and every UAV in the swarm acts as an intelligent node - capable of sensing, making decisions, and interacting with the neighboring drones. The UAV Swarm behaves as a single distributed system in which the task is shared, unlike multiple drones flying independently.

To operate drones in a swarm, communication, control, and AI-based coordination should be integrated. Each UAV should continuously measure its position, velocity, and orientation, as well as the distance to and relative position with respect to nearby UAVs. UAVs must exchange the above position and velocity information, along with the task status, with other UAVs in the team. The swarm algorithmic principles have a set of rules, viz. – Avoid collision with neighbors, match the required velocity and the direction, and stay in the group (if the task is not to be executed in a distributed manner).

Image 12.2 – SWARM of Drones[124]

Few applications of UAV Swarming are:-

a. Defense surveillance
b. Scanning large area – disaster zones
c. Disaster victim identification
d. Entertainment and public displays
e. Loitering munitions

UAV swarming applies a decentralized coordination strategy across multiple drones, enabling them to operate as a union of intelligent systems. The swarm of drones performs complex, large-scale tasks efficiently, quickly, and reliably through their internal communication and distributed control algorithms.

12.4 UAV Lightshow

One of the most attractive applications of UAVs is when a large number of drones equipped with programmable LED lights fly in precise formations to create a variety of visual patterns, symbols, and messages for people to view in the night sky, forming a coordinated aerial display. LEDs on each drone act as pixels in a three-dimensional pattern to form dynamic visual content synchronized in space and time.

UAV light show uses multiple drones flying together; however, they are fundamentally different from the functional Swarm of UAVs used in surveillance, mapping, or search and rescue. The main objective of the light show is to create a visual display and choreography of drones with pre-programmed execution, whereas a UAV swarm focuses on task execution and problem-solving with a dynamic, adaptive level of autonomy.

Image 12.3 – Intel Drone Used for Light Show[125]

For the UAV light show operation, visual patterns and animations are designed using software. Each animation frame defines the drone's 3D position and color. A safe, collision-free trajectory with timing calculations is required for all drones, along with altitude layers and spacing to create effective and attractive visuals. Each drone receives a unique flight path and LED lighting sequence before takeoff, which are executed during the show. Planning precise location, path, and movement is crucial and must be decided carefully and attentively.

Few applications of UAV Light show:-

a. Events and promotion
b. Branding, advertisement, and product launches

Image 12.4 – Drone Light Show[126]

The above image is from a Drone light show, where viewers watch a fantastic visual representation of a flying bird created by the Drone LED coordination.

UAV light shows are primarily designed for entertainment and public events and are performed with precision and synchronization. Light shows are an environmentally friendly alternative to fireworks, as they display colorful displays without creating noise or air pollution.

Benefits of AI-Enabled UAVs to society:-

a. Efficient Infrastructure Monitoring – for bridges, railways, powerlines, etc.
b. Agricultural productivity increases using AI – based crop health analysis
c. Fast disaster response and supplying aids

d. Forest and wildlife protection

e. Smart cities and traffic management

f. AI – enabled crowd analysis

Points to Remember:-

- AI and UAVs are both developed with a common aim of accomplishing tasks faster and more efficiently
- Combining both can have numerous advantages in almost any sector
- AI can be integrated into a drone for two purposes -

  1. To make the drone fly smartly through decision-making
  2. To reduce the time required to process the drone data.

Appendix

CHAPTER – WISE QUESTIONS

The chapter-wise questions are provided below to share information and spark curiosity among students. Majority of these questions are from the topics covered in this book as well as relevant to the topics, helping readers advance their learning and skills. These are sample questions and more questions can be generated as per the level of education.

Chapter 1 - Fundamentals of Unmanned Aerial Vehicles:-

1. Explain the basic concept and working principle of UAVs.
2. Differentiate between various types of UAVs with examples.
3. Classify UAVs based on international standards.
4. Compare fixed-wing and multirotor UAVs.
5. Describe hybrid UAV systems and their advantages.
6. Explain the role of each major UAV subsystem.
7. Describe power systems and battery selection criteria.
8. Explain communication systems in UAVs.
9. Discuss navigation systems used in drones.
10. Compare manual vs autonomous UAV operations.
11. Describe safety features in UAV design.
12. Explain UAV regulatory requirements in India.
13. Compare civil vs military UAV systems.
14. Describe factors affecting UAV performance.
15. What is a UAV and how does it differ from manned aircraft?

Chapter 2 - Application-Based UAV Payloads and Sensor Systems:-

1. Explain the concept of UAV payload systems.
2. Describe working of gimbal stabilization systems.
3. Write technical difference between 2-axis vs 3-axis gimbals.
4. Explain RGB camera system in UAVs.
5. Discuss applications of zoom cameras.
6. What is LiDAR and how does it work?
7. Compare the use of LiDAR and photogrammetry.
8. Describe thermal imaging camera operation.
9. Explain applications of thermal sensors.
10. Compare multispectral vs hyperspectral sensors.
11. Describe agricultural spraying systems in drones.
12. Explain aerial delivery mechanisms.
13. How does payload weight affect flight?
14. Explain defense payload systems.
15. Explain data quality in different sensors.

Chapter 3 - Fundamentals of UAV Data Acquisition and Processing:-

1. Explain UAV data acquisition process.
2. Discuss factors affecting data quality.
3. Describe data processing workflow.
4. Explain photogrammetry basics.
5. Discuss data analysis techniques in UAV mapping.
6. Explain types of UAV deliverables.
7. Compare drone data vs satellite data.
8. Discuss advantages of UAV data collection.
9. Describe ground control points (GCPs).
10. Discuss 3D mapping techniques.
11. Explain data accuracy and resolution.
12. Describe GIS integration with UAV data.
13. Explain real-time vs post-processing data.
14. Discuss applications of UAV data analysis.
15. Explain importance of data validation.

Chapter 4 - Application-Based Mission Planning for UAV Operations:-

1. What is a Ground Control Station? Describe its components.
2. Explain working of Head-Up Display (HUD).
3. Discuss steps in mission planning.
4. Explain waypoint navigation.
5. Discuss flight path optimization.
6. What is geofencing in UAVs?
7. Describe pre-flight checklist.
8. Explain risk assessment in missions.
9. Discuss weather considerations.
10. Describe failsafe mechanisms.
11. Explain autonomous vs manual missions.
12. Explain regulatory compliance in missions.
13. Discuss real-time monitoring.
14. Explain post-mission analysis.
15. Describe mission planning software tools.

Chapter 5 - UAV Applications in Surveying and Mapping:-

1. Explain UAV role in surveying.
2. Discuss road widening applications.
3. Explain the use of UAV in town planning.
4. Describe infrastructure monitoring using UAVs.
5. What is DEM and DSM?
6. Explain 3D modeling techniques.
7. Compare traditional vs UAV surveying.
8. Explain accuracy factors in mapping.
9. Explain photogrammetry in mapping.
10. Explain large area mapping challenges.
11. Describe volume calculation using drones.
12. Explain urban mapping applications.
13. Discuss legal aspects of mapping.
14. Explain real-time mapping.
15. Describe future trends in drone surveying.

Chapter 6 - UAV-Based Monitoring and Surveillance Systems:-

1. Explain UAV role in traffic management.
2. Discuss crime scene investigation using drones.
3. Explain crowd monitoring systems.
4. Describe disaster management applications.
5. Explain anti-poaching drone usage.
6. Discuss surveillance technologies.
7. Explain privacy concerns in surveillance.
8. Describe law enforcement applications.
9. Explain emergency response using UAVs.
10. Discuss border surveillance systems.
11. Explain night surveillance techniques.
12. Describe multi-drone coordination.
13. Explain communication systems in surveillance.
14. Discuss reliability and redundancy.
15. Explain role of sensors in monitoring.

Chapter 7 - UAVs for Outdoor Inspection:-

1. Explain UAV inspection concept.
2. Describe solar plant inspection.
3. Explain wind turbine inspection.
4. Discuss powerline inspection.
5. Explain transmission tower inspection.
6. Describe railway track inspection.
7. Explain chimney inspection methods.
8. Discuss inspection challenges.
9. Explain safety considerations.
10. Describe sensor selection process.
11. Explain thermal inspection.
12. Describe inspection data analysis.
13. Describe inspection workflows.
14. Explain real-time inspection.
15. Discuss future inspection trends.

Chapter 8 - UAVs for Indoor Inspection:-

1. Explain indoor UAV inspection challenges.
2. Describe boiler inspection.
3. Explain chimney inspection indoors.
4. Discuss storage tank inspection.
5. Explain sewer inspection.
6. Describe tunnel inspection.
7. Explain navigation without GPS.
8. Discuss collision avoidance systems.
9. Describe confined space challenges.
10. Explain sensor selection.
11. Discuss SLAM technology.
12. Explain data capture methods.
13. Describe safety precautions and operational constraints.
14. Explain inspection drones design.
15. Explain real-time monitoring.

Chapter 9 - Role of UAVs in Smart Farming:-

1. Explain UAV role in agriculture.
2. Describe spraying drones.
3. Explain spreading drones.
4. Discuss crop monitoring.
5. Explain NDVI analysis.
6. Describe precision farming.
7. Explain pest detection.
8. Discuss irrigation monitoring.
9. Explain yield estimation.
10. Describe farm mapping.
11. Explain multispectral sensors use.
12. Discuss benefits to farmers.
13. Explain challenges in adoption.
14. Explain automation in farming.
15. Explain data-driven agriculture.

Chapter10 - Entertainment and Professional Use of UAV:-

1. Explain UAV role in photography.
2. Describe videography applications.
3. Explain drone racing.
4. Discuss leisure drone use.
5. Explain drone soccer concept.
6. Explain media and advertising use.
7. Discuss business opportunities.
8. Explain legal regulations.
9. Describe live event coverage.
10. Discuss safety in entertainment drones.
11. Explain FPV systems.
12. Describe challenges in industry.
13. Explain equipment selection.
14. How drones can help in product launch?
15. What are the parts of FPV drones built for bird's eye view?

Chapter 11 - Role of UAVs in Modern Defence:-

1. Explain UAV role in defense sector.
2. Discuss generations of warfare.
3. Explain evolution of UAVs.
4. Describe surveillance drones.
5. Explain combat UAVs.
6. Discuss intelligence gathering.
7. Explain UAV swarms in defense.
8. Explain communication systems.
9. Discuss ethical concerns.
10. Describe counter-drone systems.
11. Explain stealth technology.
12. Discuss mission planning in defense.
13. Explain electronic warfare.
14. Discuss challenges in defense UAVs.
15. Explain future military UAVs.

Chapter 12 - Artificial Intelligence in UAV Systems:-

1. Explain AI in UAVs.
2. Discuss object counting.
3. Explain target detection.
4. Describe tracking systems.
5. Explain computer vision basics.
6. Discuss machine learning models.
7. Explain UAV swarming.
8. Describe autonomous navigation.
9. Explain path planning algorithms.
10. Discuss real-time decision making.
11. Explain drone light shows.
12. Explain AI-based safety systems.
13. Discuss future AI trends.
14. Explain limitations of AI in UAVs.
15. Describe ethical considerations.

References

1 https://www.jkproducts.co.za/products/hangar-9-j-3-cub-10cc-arf-82-5?srsltid=AfmBOoqTTfmxWQTQqohq83JSB2mrmuU0PAbXJd4dHtn3x-RnqJZgdGNs

2 https://acecoretechnologies.com/noa/

3 https://dronelitic.com/threod-systems-eos-mini-vtol-uas/

4 161012-N-JE250-047 by Office of Naval Research is licensed under CC BY 2.0 https://www.flickr.com/photos/usnavyresearch/29669467064/

5 https://rcworld.us/rc-bird-eagle-ornithopter-flapping-wing-2-4ghz-rtf/

6 Pandya, Garvit. (2021). Basics of Unmanned Aerial Vehicles (Time to start working on Drone Technology).

7 https://dgca.gov.in/digigov-portal/jsp/dgca/homePage/viewPDF.jsp?page=InventoryList/headerblock/drones/D3X-X1.pdf

8 https://hazmatresource.com/product/unmanned-systems/unmanned-aerial-systems/flir-black-hornet-3-nano-drone/?srsltid=AfmBOorSW2-e6mUs8p7NV5DbmXVEkxAzBo8Bq5M-cirxVUjldMh_EGnJ

9 https://dronarbutiken.se/products/parrot-anafi-ukr?srsltid=AfmBOopCdmztcN142tveb-0KV3uPQwKgYbmuMKVzKS-bFKfX6eIr2qbv

10 https://www.altiuas.com/transition/

11 https://velos-rotors.com/velos-v3/

12 https://www.eurasiantimes.com/us-deployed-rq-170-stealth-uav-to-spy-on-russian-military/

13 https://www.dubro.com/products/raider-rc-plane-kit-from-old-school-model-works?srsltid=AfmBOor7XbeA72ONjSbptw3JEK8dP4BeKbc_Np4eWE8C0VeT5TbWmJc5

14 https://www.unmannedsystemstechnology.com/feature/inrunner-bldc-motors-for-uav-applications/

15 https://www.unmannedsystemstechnology.com/expo/2-stroke-uav-engines/

16 https://store.micropilot.com/product-p/a-2028g2.htm

17 https://www.3dxr.co.uk/autopilots-c2/the-cube-aka-pixhawk-2-1-c9/here-gnss-and-rtk-c11/cubepilot-here4-blue-multiband-rtk-gps-with-bluetooth-5-2-hx4-06279-p5742

18 https://www.unmannedsystemstechnology.com/expo/drone-and-uav-remote-controllers/

[19] Pandya, Garvit. (2021). Basics of Unmanned Aerial Vehicles (Time to start working on Drone Technology).

[20] https://gremsy.com/?srsltid=AfmBOoo7gQQ9LHBTw9J2AmsnqYb9-Bfn2U158j2Y3YPniBjtB_HvBjFa

[21] https://www.bhphotovideo.com/c/product/1172251-REG/sony_alpha_a6000_mirrorless_digital.html

[22] https://www.adti.camera/

[23] https://www.viewprouav.com/product/q10f-10x-zoom-camera.html

[24] https://autonomoustuff.com/products/ouster-os0

[25] https://www.skydroid.xin/#/productCenter/productInfo?id=63&sort=10

[26] https://www.ebay.com/itm/235821607082

[27] https://ageagle.com/drone-sensors/rededge-p-high-res-multispectral-camera/

[28] https://ageagle.com/drone-sensors/rededge-p-high-res-multispectral-camera/

[29] Jasinski, Janusz & Pietrek, Slawomir & Walczykowski, Piotr & Orych, Agata. (2010). Acquisition of spectral reflectance characteristics of land cover features based on hyperspectral images

[30] https://www.specim.com/afx/

[31] https://spectrum.ieee.org/test-run-confirms-that-drones-can-safely-deliver-organs-for-transplant

[32] https://zipline.imagerelay.comsb6f461343-ecd8-4264-a842-ecd4f395bf98zipline-assets

[33] https://www.trilliumeng.com/gimbals/hd95-mvs

[34] https://www.the-sun.com/wp-content/uploads/sites/6/2021/08/NINTCHDBPICT000231859399.jpg

[35] https://www.pix-pro.com/blog/drone-mapping

[36] https://dronesafestore.com/blog/drone-industry-products

[37] https://help.dronedeploy.com/hc/en-us/articles/1500004965282-What-are-Ground-Control-Points-GCPs-Checkpoints

[38] https://wingtra.com/ground-control-points-how-many-do-you-need-and-when-are-checkpoints-enough/

[39] Pepe, M.; Alfio, V.S.; Costantino, D. UAV Platforms and the SfM-MVS Approach in the 3D Surveys and Modelling: A Review in the Cultural Heritage Field. *Appl. Sci.* **2022**, *12*, 12886. https://doi.org/10.3390/app122412886

[40] https://pixelement.com/blog/2024/06/17/importance-of-imagery-overlap.html

[41] https://www.dronedeploy.com/

[42] https://www.pix4d.com/product/pix4dmatic-large-scale-photogrammetry-software/

[43] https://www.pix4d.com/product/pix4dmapper-photogrammetry-software/

[44] https://qgis.org/

[45] https://www.agisoft.com/

46 https://www.bluemarblegeo.com/global-mapper/

47 https://www.propelleraero.com/

48 https://opendronemap.org/

49 https://www.bentley.com/software/reality-and-spatial-modeling/

50 https://www.esri.com/en-us/arcgis/products/arcgis-reality/products/site-scan-for-arcgis

51 https://www.simactive.com/

52 https://matthewjonesphotography.co.uk/drone-surveying/

53 https://raad.com/blog/lidar-vs-photogrammetry-2025

54 https://www.cyclomedia.com/us/what-is-point-cloud-data

55 https://www.researchgate.net/publication/346902099_Digital_Elevation_Models

56 https://gisrsstudy.com/digital-elevation-model/

57 https://www.fathom.global/academic-papers/a-30-m-global-map-of-elevation-with-forests-and-buildings-removed/

58 https://www.luxcarta.com/blog/dem-dtm-dsm

59 https://docs.up42.com/data/elevation-models

60 https://atlas.co/images/posts/contour-lines.png

61 https://gis.stackexchange.com/questions/262539/calculating-slope-and-aspect-from-dtm-lidar-elevation-data-in-arcgis-desktop

62 https://community.esri.com/t5/arcgis-drone2map-questions/drone2map-thermal-orthomosaic-using-a-dji-matrice/td-p/1321878

63 https://tmt.co.il/product/solar-panels-thermal-visual-inspection/

64 https://climatedataguide.ucar.edu/climate-data/ndvi-normalized-difference-vegetation-index-3rd-generation-nasagfsc-gimms

65 https://www.esri.com/arcgis-blog/products/spatial-analyst/analytics/getting-the-most-out-of-zonal-statistics

66 https://labelyourdata.com/articles/satelite-vs-drones-imagery#:~:text=Analyzing%20drone%20imagery%20can%20be,effort%20required%20for%20data%20processing. https://labelyourdata.com/articles/satelite-vs-drones-imagery

67 https://www.hammermissions.com/post/how-to-3d-model-a-cell-tower-for-inspection-a-step-by-step-guide

68 https://www.autelpilot.com/blogs/news/how-are-drones-used-for-road-construction-and-surveying

69 https://www.autelpilot.com/blogs/faq/the-role-of-drones-in-urban-planning-and-architecture

70 https://dronephotographyservices.co.uk/mapping/3d-mapping/

71 https://geospatial.chcnav.com/products/chcnav-C30

72 https://www.hammermissions.com/post/how-to-capture-data-for-3d-mapping-using-drones

73 https://fotokite.com/public-safety-uas/traffic-management-drone/

74 https://www.pix4d.com/blog/accident-and-crime-scene-investigation/

75 https://agiindia.com/life-saving-insights-in-disaster-zones-with-drones/

76 https://becominghuman.ai/ai-powered-drones-are-being-sent-to-africa-to-catch-the-poachers-1378ec17bc5d

77 https://enterprise.news/climate/en/news/story/4af99887-051c-48c9-b8f8-7af8291db914/norway%E2%80%99s-scatec-signs-an-agreement-to-build-a-1-gw-solar-farm-in-egypt

78 https://shop.flyindiamonds.com/siyi-zt6-mini/

79 Photo by Kelly : https://www.pexels.com/photo/windmills-in-agricultural-field-against-cloudy-sky-4394353/

80 https://www.hammermissions.com/post/wind-turbine-inspections-using-hammer-missions-on-android

81 https://thedronelifenj.com/linemens-guide-drone-inspections-transmission-line/

82 https://thedronelifenj.com/linemens-guide-drone-inspections-transmission-line/

83 https://thedronelifenj.com/linemens-guide-drone-inspections-transmission-line/

84 Photo by Tom Fisk: https://www.pexels.com/photo/aerial-shot-of-railroad-3947562/

85 Jia, Lei & Zhu, Ming & Sherman, Ryan & Park, Jeewoong & Jiang, Yingtao & Teng, Hualiang. (2019). Rail Defect Detection Technology: A Review of the Current Methods and an Acoustic Based Method Proposed for High-Speed-Rail.

86 https://www.railway-technology.com/news/360-rail-services-taps-ardenna-for-drone-based-rail-yard-inspection/?cf-view

87 Photo by Bl4ke : https://www.pexels.com/photo/birds-eye-view-of-industrial-chimney-19784122/

88 https://www.aetha.global/product/flybotix-asio-x

89 https://www.ascentboilers.com/industrial-steam-boiler-for-power-plants-9127024.html

90 https://www.petrosync.com/blog/storage-tank/

91 https://coptrz.com/blog/drones-for-internal-inspection-access-all-areas/

92 https://fly4future.com/development-and-prototyping/drones-for-inspection-of-highway-tunnels/

93 https://www.jmrdrone.com/agriculture-drone-parts/62412664.html

94 https://mavdrones.com/product/buy-this-hobbywing-brushless-water-pump-5l/?gad_source=1&gclid=CjwKCAjwps-zBhAiEiwALwsVYRWwa2M3vOTLHU04l-jZiiHVEOKsH_E1QHe-aAOn6NIRXXi_in935BoC6r8QAvD_BwE

95 https://rcdrone.top/products/eft-drone-flow-meter-liquid-measurement

96 https://www.xpneumatic.com/pneumatic-fittings-2/plastic-pneumatic-fitting/

97 https://www.amazon.com/KOSDFOGE-Agricultural-Protection-Spraying-Atomization/dp/B0C8PM7GTP

98 https://titanflying.en.made-in-china.com/product/UwBaNFCDQRhX/China-New-Agricultural-Drone-Uav-High-Pressure-Atomizer-Nozzle-Extension-for-Outer-Diameter-6mm-8mm-Hose-Nozzle-for-Hobbywing-X8.html

99 https://droneag.farm/solutions/dji-agras-spraying-drone/

100 https://www.effort-tech.com/en/bs

101 https://vurimiaiglobal.com/book-drone-spray-services/

102 https://gisgeography.com/ndvi-normalized-difference-vegetation-index/

103 https://geprc.com/product/geprc-moz7-v2-analog-long-range-fpv-drone/

104 https://www.rotorama.com/product/runcam-phoenix-2-nite-kamera

105 https://drone24hours.com/en/products/eachine-ev800d?srsltid=AfmBOopqjThahIMXWWvaPb1jpOoz7B3_DGZB00Nb3Kvjwx-JWX_0f6Gs

106 https://thefpvproject.com/product/caddx-infra-v2-fpv-camera-kit/?srsltid=AfmBOopNwlsAMPpegFL92krGZvQm8RRffqm3y-r8wXDtSv4hRlD8BZPj

107 https://www.rotorama.com/product/walksnail-avatar-hd-goggles-x

108 https://www.multigp.com/multigp-tracks/?trackId=117

109 https://reference.tlciscreative.com/private/technology-meets-creativity/air-soccer/

110 https://dronesoccer.or.kr/

111 https://dronesoccer.or.kr/

112 https://dras.in/evolution-of-warfare-with-technology/

113 File:556th RS AQM-34 Drone.jpg, https://commons.wikimedia.org/w/index.php?title=File:556th_RS_AQM-34_Drone.jpg&oldid=910804842 (last visited March 11, 2026).

114 https://www.robinradar.com/blog/evolution-of-the-drone-threat-part-2

115 *Unmanned Aerial Vehicles: Robotic Air Warfare 1917-2007*, By Steven Zaloga, Osprey Publishing, 19 July 2011, page 22

116 By Lt. Col. Leslie Pratt - MQ-9 Reaper UAV.jpg, Public Domain, https://commons.wikimedia.org/w/index.php?curid=68095681

117 https://en.wikipedia.org/wiki/General_Atomics_MQ-9_Reaper

118 https://www.autonomyglobal.co/ai-powered-hermes-900-hunts-iranian-missile-launchers-in-24-7-u-s-israeli-operation/

119 https://en.wikipedia.org/wiki/Elbit_Hermes_900

120 https://www.defendtex.com/uav/

121 https://www.defendtex.com/wp-content/uploads/DTX_D40_Payload_Datasheet-1.pdf

122 https://www.griffaviation.com/products/griff-60

123 https://solvi.ag/features/plant-counts

124 https://www.defensenews.com/unmanned/2022/03/01/us-army-to-demo-offensive-drone-swarms-in-next-project-convergence/

125 https://www.commercialuavnews.com/infrastructure/intels-drone-light-shows-will-impact-creation-enterprise-fleets

126 https://www.whidbeynewstimes.com/wp-content/uploads/2024/07/36807984_web1_Drones2.jpg

amcontent.com/pod-product-compliance
Source LLC
TN
134210926
LV00006B/307
* 9 8 9 0 3 9 2 7 4 2 5 *